MIND AND MATTER

Erwin Schrödinger

CHAPTER I

The Physical Basis of Consciousness

THE PROBLEM

The world is a construct of our sensations, *perceptions, memories. It is convenient to regard it as existing objectively on its own. But it certainly does not become manifest by its mere existence. It's becoming manifest is conditional* on very special goings-on in very special parts of this very world, namely on certain events that happen in a brain. That is an inordinately peculiar kind of implication, which prompts the question: What particular properties distinguish these brain processes and enable them to produce the manifestation? Can we guess which material processes have this power, which not? Or simpler: What kind of material process is directly associated with consciousness?

A rationalist may be inclined to deal curtly with this

question, roughly as follows. From our own experience, and as regards the higher animals from analogy, consciousness is linked up with certain kinds of events in organized, living matter, namely, with certain nervous functions. How far back or 'down' in the animal kingdom there is still some sort of consciousness, and what it may be like in its early stages, are gratuitous speculations, questions that cannot be answered and which ought to be left to idle dreamers. It is still more gratuitous to indulge in thoughts about whether perhaps other events as well, events in inorganic matter, let alone all material events, are in some way or other associated with consciousness. All this is pure fantasy, as irrefutable as it is unprovable, and thus of no value for knowledge.

He who accepts this brushing aside of the question ought to be told what an uncanny gap he thereby allows to remain in his picture of the world. For the turning-up of nerve cells and brains within certain strains of organisms is a very special event whose meaning and significance is quite well under- stood. It is a special kind of mechanism by which the individual responds to alternative situations by accordingly alternating behaviour, a mechanism for adaptation to a changing surrounding. It is the most elaborate and the most ingenious among all such mechanisms, and wherever it turns up it rapidly gains a dominating role. However, it is not *sui generis*. Large groups of organisms, in particular the plants, achieve very similar performances in an entirely different fashion.

Are we prepared to believe that this very special turn in the development of the higher animals, a turn that might after all have failed to appear, was a necessary condition for the world to flash up to itself in the light of consciousness? Would it otherwise have remained a play before empty benches, not existing for anybody, thus quite properly speaking not existing? This would seem to me the bankruptcy of a world picture. The urge to find a way out of this impasse ought not to be damped by the fear of incurring the wise rationalists' mockery.

According to Spinoza every particular thing or being is a modification of the infinite substance, i.e. of God. It expresses itself by each of his attributes, in particular that of extension and that of thought. The first is its bodily existence in space and time, the second is - in the case of a living man or animal - his mind. But to Spinoza any inanimate bodily thing is at the same time also 'a thought of God', that is, it exists in the second attribute as well. We encounter here the bold thought of universal animation, though not for the first time, not even in Western philosophy. Two thousand years earlier the Ionian philosophers acquired from it the surname of *hylo;:,oists*. After Spinoza the genius of Gustav Theodor Fechner did not shy at attributing a soul to a plant, to the earth as a celestial body, to the planetary system, etc. I do not fall in with these fantasies, yet I should not like to have to pass judgment as to who has come nearer to the deepest truth, Fechner or the bankrupts of rationalism.

A TENTATIVE ANSWER

You see that all the attempts at extending the domain of consciousness, asking oneself whether anything of the sort might be reasonably associated with other than nervous processes, needs must run into unproved and unprovable speculation. But we tread on firmer ground when we start in the opposite direction. Not every nervous process, nay by no means every cerebral process, is accompanied by conscious- ness. Many of them are not, even though physiologically and biologically they are very much like the 'conscious' ones, both in frequently consisting of afferent impulses followed by efferent ones, and in their biological significance of regulating and timing reactions partly inside the system, partly towards a changing environment. **In** the first instance we meet here with the reflex actions in the vertebral ganglia and in that part of the nervous system which they control. But also (and this we shall make our special study) many reflexive processes exist that do pass through the brain, yet do not fall into consciousness at all or have very nearly ceased to do so. For in the latter case the distinction is not sharp; intermediate degrees between fully conscious and completely unconscious occur. By examining various representatives of physiologically very similar processes, all playing within our own body, it ought not to be too difficult to find out by observation and reasoning the distinctive characteristics we are looking for.

To my mind the key is to be found in the following

well-known facts. Any succession of events in which we take part with sensations, perceptions and possibly with actions gradually drops out of the domain of consciousness when the same string of events repeats itself in the same way very often. But it is immediately shot up into the conscious region, if at such a repetition either the occasion or the environmental conditions met with on its pursuit differ from what they were on all the previous incidences. Even so, at first anyhow, only those modifications or 'differentials' intrude into the conscious sphere that distinguish the new incidence from previous ones and thereby usually call for 'new considerations'. Of all this each of us can supply dozens of examples out of personal expenence, so that I may forgo enumerating any at the moment.

The gradual fading from consciousness is of outstanding importance to the entire structure of our mental life, which is wholly based on the process of acquiring practice by repetition, a process which Richard Semon has generalized to the concept of *Mneme,* about which we shall have more to say later. A single experience that is never to repeat itself is biologically irrelevant. Biological value lies only in learning the suitable reaction to a situation that offers itself again and again, in many cases periodically, and always requires the same response if the organism is to hold its ground. Now from our own inner experience we know the following. On the first few repetitions a new element turns up in the mind, the 'already met with' or 'notal' as Richard Avenarius has called it. On frequent repetition the whole string of events becomes more and more of a routine, it becomes more and more uninteresting, the responses become ever more reliable according as they fade from consciousness. The boy recites his poem, the girl plays her piano sonata 'well-nigh in their sleep'. We follow the habitual path to our workshop, cross the road at the customary places, turn into side-streets, etc., whilst our thoughts are occupied with entirely different things. But whenever the situation exhibits a relevant differential - let us say the road is up at the place where we used to cross it, so that we have to make a detour - this differential and our response to it intrude into consciousness, from which, however, they soon fade below the threshold, if the differential becomes a constantly repeated feature. Faced with changing alternatives, bifurcations develop and may be fixed in the same way. We branch off to the University Lecture Rooms or to the Physics Laboratory at the right point without much thinking, provided that both are frequently occurring destinations.

Now in this fashion differentials, variants of response, bifur- cations, etc., are piled up one upon the other in unsurveyable abundance, but only the most recent ones remain in the domain of consciousness, only those with regard to which the living substance is still in the stage of learning or practising. One might say, metaphorically, that consciousness is the tutor who supervises the education of the living substance, but leaves his pupil alone to deal with all those tasks for which he is already sufficiently trained. But I wish to underline three times in red ink that I mean this only as a metaphor. The fact is only this, that new situations and the new responses they prompt are kept in the light of consciousness; old and well practised ones are no longer so.

Hundreds and hundreds of manipulations and performances

of everyday life had all to be learnt once, and that with great attentiveness and painstaking care. Take for example a small child's first attempts in walking. They are eminently in the focus of consciousness; the first successes are hailed by the peformer with shouts of joy. When the adult laces his boots, switches on the light, takes off his clothes in the evening, eats with knife and fork ... , these performances, that all had to be toilsomely learnt, do not in the least disturb him in the thoughts in which he may just be engaged. This may occasionally result in comical miscarriages. There is the story of a famous mathematician, whose wife is said to have found him lying in his bed, the lights switched off, shortly after an invited evening party had gathered in his house. What had happened? He had gone to his bedroom to put on a fresh shirt-collar. But the mere action of taking off the old collar had released in the man, deeply entrenched in thought, the string of performances that habitually followed in its wake.

Now this whole state of affairs, so well known from the

ontogeny of our mental life, seems to me to shed light on the *phylogeny* of unconscious nervous processes, as in the heart beat, the peristalsis of the bowels, etc. Faced with nearly constant or regularly changing situations, they are very well and reliably practised and have, therefore, long ago dropped from the sphere of consciousness. Here too we find intermediate grades, for example, breathing, that usually goes on inadvertently, but may on account of differentials in the situation, say in smoky air or in an attack of asthma, become modified and conscious. Another instance is the bursting into tears for sorrow, joy or bodily pain, an event which, though conscious, can hardly be influenced by will. Also comical miscarriages of a mnemically inherited nature occur, as the bristling of the hair by terror, the ceasing of secretion of saliva on intense excitement, responses which must have had some significance in the past, but have lost it in the case of man.

I doubt whether everybody will readily agree with the next step, which consists in extending these notions to other than nervous processes. For the moment I shall only briefly hint at it, though to me personally it is the most important one. For this generalization precisely sheds light on the problem from which we started: What material events are associated with, or accompanied by, consciousness, what not? The answer that I suggest is as follows: What in the preceding we have said and shown to be a property of nervous processes is a property of organic processes in general, namely, to be associated with consciousness inasmuch as they are new.

In the notion and terminology of Richard Semon the ontogeny not only of the brain but of the whole individual soma is the 'well memorized' repetition of a string of events that have taken place in much the same fashion a thousand times before. I ts first stages, as we know from our own experience, are unconscious - first in the mother's womb; but even the ensuing weeks and months of life are for the greatest part passed in sleep. During this time the infant carries on an evolution of old standing and habit, in which it meets with conditions that from case to case vary very little. The ensuing organic development begins to be accompanied by consciousness only inasmuch as there are organs that gradually take up interaction with the environment, adapt their functions to the changes in the situation, are influenced, undergo practice, are in special ways modified by the surroundings. We higher vertebrates possess such an organ mainly in our nervous system. Therefore consciousness is associated with those of its functions that adapt themselves by what we call experience to a changing environment. The nervous system is the place where our species is still engaged in phylogenetic transformation; metaphorically speaking it is the 'vegetation top' (*Vegetationsspit::::,e)* of our stem. I would summarize my general hypothesis thus: consciousness is associated with the *learning* of the living substance; its *knowing how (Konnen)* is uncon- sc1Ous.

ETHICS

Even without this last generalization, which to me is very important but may still seem rather dubious to others, the theory of consciousness that I have adumbrated seems to pave the way towards a scientific understanding of ethics.

At all epochs and with all peoples the background of every ethical code (*Tugendlehre)* to be taken seriously has been, and is, self-denial *(Selbstuberwindung)*. The teaching of ethics always assumes the form of a demand, a challenge, of a 'thou shalt', that is in some way opposed to our primitive will. Whence comes this peculiar contrast between the 'I will' and the 'thou shalt'? Is it not absurd that I am supposed to suppress my primitive appetites, disown my true self, be different from what I really am? Indeed in our days, more perhaps than in others, we hear this demand often enough mocked at. 'I am as I am, give room to my individuality! Free development to the desires that nature has planted in me! All the shalls that oppose me in this are nonsense, priests' fraud. God is Nature, and Nature may be credited with having formed me as she wants me to be.' Such slogans are heard occasionally. It is not easy to refute their plain and brutal obviousness. Kant's imperative is avowedly irrational.

But fortunately the scientific foundation of these slogans is worm-eaten. Our insight into the 'becoming' *(das Werden)* of the organisms makes it easy to understand that our conscious life - I will not say shall be, but that it actually is necessarily a continued fight against our primitive ego. For our natural self, our primitive will with its innate desires, is obviously the mental correlate of the material bequest received from our ancestors. Now as a species we are developing, and we march in the front-line of generations; thus every day of a man's life represents a small bit of the evolution of our species, which is still in full swing. It is true that a single day of one's life, nay even any individual life as a whole, is but a minute blow of the chisel at the ever unfinished statue. But the whole enormous evolution we have gone through in the past, it too has been brought about by myriads of such minute chisel blows. The material for this transformation, the presupposition for its taking place, are of course the inheritable spontaneous muta- tions. However, for selection among them, the behaviour of the carrier of the mutation, his habits of life, are of outstanding importance and decisive influence. Otherwise the origin of species, the ostensibly directed trends along which selection proceeds, could not be understood even in the long spaces of time which are after all limited and whose limits we know quite well.

And thus at every step, on every day of our life, as it were,

something of the shape that we possessed until then has to change, to be overcome, to be deleted and replaced by something new. The resistance of our primitive will is the psychical correlate of the resistance of the existing shape to the transforming chisel. For we ourselves are chisel and statue, conquerors and conquered at the same time - it is a true continued 'self-conquering' *(Selbstiiberwindung)*.

But is it not absurd to suggest that this process of evolution should directly and significantly fall into consciousness, con- sidering its inordinate slowness not only compared with the short span of an individual life, but even with historical epochs? Does it not just run along unnoticed?

No. In the light of our previous considerations this is not so.

They culminated in regarding consciousness as associated with such physiological goings-on as are still being trans- formed by mutual interaction with a changing environment. Moreover, we concluded that only those modifications become conscious which are still in the stage of being trained, until, in a much later time, they become a hereditarily fixed, well-trained and unconscious possession of the species. In brief: consciousness is a phenomenon in the zone of evolution. This world lights up to itself only where or only inasmuch as it develops, procreates new forms. Places of stagnancy slip from consciousness; they may only appear in their interplay with places of evolution.

If this is granted it follows that consciousness and discord with one's own self are inseparably linked up, even that they must, as it were, be proportional to each other. This sounds a paradox, but the wisest ofall times and peoples have testified to confirm it. Men and women for whom this world was lit in an unusually bright light of awareness, and who by life and word have, more than others, formed and transformed that work of art which we call humanity, testify by speech and writing or even by their very lives that more than others have they been torn by the pangs of inner discord. Let this be a consolation to him who also suffers from it. Without it nothing enduring has ever been begotten.

Please do not misunderstand me. I am a scientist, not a teacher of morals. Do not take it that I wish to propose the idea of our species developing towards a higher goal as an effective motive to propagate the moral code. This it cannot be, since it is an unselfish goal, a disinterested motive, and thus, to be accepted, already presupposes virtuousness. I feel as unable as anybody else to explain the 'shall' of Kant's imperative. The ethical law in its simplest general form (be unselfish!) is plainly a fact, it is there, it is agreed upon even by the vast majority of those who do not very often keep it. I regard its puzzling existence as an indication of our being in the beginning of a biological transformation from an egoistic to an altruistic general attitude, of man being about to become an *animal social.* For a solitary animal egoism is a virtue that tends to preserve and improve the species; in any kind of community it becomes a destructive vice. An animal that embarks on forming states without greatly restricting egoism will perish. Phylogenetically much older state-formers as the bees, ants and termites have given up egoism completely. However, its next stage, national egoism or briefly nationalism, is still in full swing with them. A worker bee that goes astray to the wrong hive is murdered without hesitation.

Now in man something is, so it seems, on the way that is not infrequent. Above the first modification clear traces of a second one in similar direction are noticeable long before the first is even nearly achieved. Though we are still pretty vigorous egoists, many of us begin to see that nationalism too is a vice that ought to be given up. Here perhaps something very strange may make its appearance. The second step, the pacification of the struggle of peoples, may be facilitated by the fact that the first step is far from being achieved, so that egoistic motives still have a vigorous appeal. Each one of us is threatened by the terrific new weapons of aggression and is thus induced to long for peace among the nations. If we were bees, ants or Lacedaemonian warriors, to whom personal fear does not exist and cowardice is the most shameful thing in the world, warring would go on for ever. But luckily we are only men - and cowards.

The considerations and conclusions of this chapter are, with me, of very old standing; they date back more than thirty years. I never lost sight of them, but I was seriously afraid that they might have to be rejected on the ground that they appear to be based on the 'inheritance of acquired characters', in other words on Lamarckism. This we are not inclined to accept. Yet even when rejecting the inheritance of acquired characters, in other words accepting Darwin's Theory of Evolution, we find the behaviour of the individuals of a species having a very significant influence on the trend of evolution, and thus feigning a sort of sham-Lamarckism. This is explained, and clinched by the authority of Julian Huxley, in the following chapter, which, however, was written with a slightly different problem in view, and not just to lend support to the ideas put forward above.

CHAPTER 2

The Future of Understanding1

A BIOLOGICAL BLIND ALLEY?

Wemay, I believe, regard it as extremely improbable that our understanding of the world represents any definite or final stage, a maximum or optimum in any respect. By this I do not mean merely that the continuation of our research in the various sciences, our philosophical studies and religious endeavour are likely to enhance and improve our present outlook. What we are likely to gain in this way in the next, say, two and a half millennia - estimating from what we have gained since Protagoras, Democritus and Antisthenes - is insignificant compared with what I am here alluding to. There is no reason whatever for believing that our brain is the

supreme *ne plus ultra* of an organ of thought in which the world is reflected. It is more likely than not that a species could acquire a similar contraption whose corresponding imagery compares with ours as ours with that of the dog, or his in turn with that of a snail.

If this be so, then - though it is not relevant in principle - it interests us, as it were for personal reasons, whether anything of the sort could be reached on our globe by our own offspring or the offspring of some of us. The globe is all right. It is a fine young leasehold, still to run under acceptable conditions of living for about the time it took us (say 1,000 million years) to develop from the earliest beginnings into what we are now.

But are we ourselves all right? If one accepts the present theory of evolution - and we have no better - it might seem that we have been very nearly cut off from future develop- ment. Is there still physical evolution to be expected in man, I mean to say relevant changes in our physique that become gradually fixed as inherited features, just as our present bodily self is fixed by inheritance - genotypical changes, to use the technical term of the biologist? This question is difficult to answer. We may be approaching the end of a blind alley, we may even have reached it. This would not be an exceptional event and it would not mean that our species would have to become extinct very soon. From the geological records we know that some species or even large groups seem to have reached the end of their evolutionary possibilities a very long time ago, yet they have not died out, but have remained unchanged, or without significant change, for many millions of years. The tortoises, for instance, and the crocodiles are in this sense very old groups, relics of a far remote past; we are also told that the whole large group of insects are more or less in the same boat - and they comprise a greater number of separate species than all the rest of the animal kingdom taken together. But they have changed very little in millions of years, while the rest of the living surface of the earth has during this time undergone change beyond recognition. What barred further evolution in the insects was probably this - that they had adopted the plan (you will not misunderstand this figurative expression) - that they had adopted the plan of wearing their skeleton outside instead of inside as we do. Such an outside armour, while affording protection in addition to mechanical stability, cannot grow as the bones of a mammal do between birth and maturity. This circumstance is bound to render gradual adaptive changes in the life-history of the individual very difficult.

In the case of man several arguments seem to militate against

further evolution. The spontaneous inheritable changes - now called mutations - from which, according to Darwin's theory, the 'profitable' ones are automatically selected, are as a rule only small evolutionary steps, affording, if any, only a slight advantage. That is why in Darwin's deductions an important part is attributed to the usually enormous abundance of offspring, of which only a very small fraction can possibly survive. For only thus does a small amelioration in the chance of survival seem to have a reasonable likelihood of being realized. This whole mechanism appears to be blocked in civilized man - in some respects even reversed. We are, generally speaking, not willing to see our fellow-creatures suffer and perish, and so we have gradually introduced legal and social institutions which on the one hand protect life, condemn systematic infanticide, try to help every sick or frail human being to survive, while on the other hand they have to replace the natural elimination of the less fit by keeping the offspring within the limits of the available livelihood. This is achieved partly in a direct way, by birth control, partly by preventing a considerable proportion of females from mating. Occasionally - as this generation knows all too well - the insanity of war and all the disasters and blunders that follow in its wake contribute their share to the balance. Millions of adults and children of both sexes are killed by starvation, exposure, epidemics. While in the far remote past warfare between small tribes or clans is supposed to have had a positive selection value, it seems doubtful whether it ever had in historical times, and doubtless war at present has none. It means an indiscriminate killing, just as the advances in medicine and surgery result in an indiscriminate saving of lives. While justly and diametrically opposite in our esteem, both war and medical art seem to be of no selection value whatever.

These considerations suggest that as a developing species we have come to a standstill and have little prospect of further biological advance. Even if this were so, it need not bother us. We might survive without any biological change for millions of years, like the crocodiles and many insects. Still from a certain philosophical point of view the idea is depressing, and I should like to try and make out a case for the contrary. To do so I must enter on a certain aspect of the theory of evolution which I find supported in Professor Julian Huxley's well- known book on Evolution, [1] an aspect which according to him is not always sufficiently appreciated by recent evolutionists.

Popular expositions of Darwin's theory are apt to lead you to a gloomy and discouraging view on account of the apparent passivity of the organism in the process of evolution. Muta- tions occur spontaneously in the genom - the 'hereditary substance'. We have reason to believe that they are mainly due to what the physicist calls a thermodynamic fluctuation - in other words to pure chance. The individual has not the slightest influence on the hereditary treasure it receives from its parents, nor on the one it leaves to its offspring. Mutations that occur are acted on by 'natural selection of the fittest'. This again seems to mean pure chance, since it means that a favourable mutation increases the prospect for the individual of survival and of begetting offspring, to which it transmits the mutation in question. Apart from this, its activity during its lifetime seems to be biologically irrelevant. For nothing of it has an influence on the offspring: acquired properties are not inherited. Any skill or training attained is lost, it leaves no trace, it dies with the individual, it is not transmitted. An intelligent being in this situation would find that nature, as it were, refuses his collaboration - she does all herself, dooms the individual to inactivity, indeed to nihilism.

As you know, Darwin's theory was not the first systematic theory of evolution. It was preceded by the theory of Lamarck, which rests entirely on the assumption that any new features an individual has acquired by specific surroundings or beha- viour during its lifetime before procreation can be, and usually are, passed on to its progeny, if not entirely, at least in traces. Thus if an animal by living on rocky or sandy soil produced protecting calluses on the soles of its feet, this callosity would gradually become hereditary so that later generations would receive it as a free gift without the hardship of acquiring it. In the same way the strength or skill or even substantial adaptation produced in any organ by its being continually used for certain ends would not be lost, but passed on, at least partly, to the offspring. This view not only affords a very simple understanding of the amazingly elaborate and specific adaptation to environment which is so characteristic of all living creatures. It is also beautiful, elating, encouraging and invigorating. It is infinitely more attractive than the gloomy aspect of passivity apparently offered by Darwinism. An intelligent being which considers itself a link in the long chain of evolution may, under Lamarck's theory, be confident that its striving and efforts for improving its abilities, both bodily and mental, are not lost in the biological sense but form a small but integrating part of the striving of the species towards higher and ever higher perfection.

Unhappily Lamarckism is untenable. The fundamental assumption on which it rests, namely, that acquired properties can be inherited, is wrong. To the best of our knowledge they cannot. The single steps of evolution are those spontaneous and fortuitous mutations which have nothing to do with the behaviour of the individual during its lifetime. And so we appear to be thrown back on the gloomy aspect of Darwinism that I have depicted above.

BEHAVIOUR INFLUENCES SELECTION

I now wish to show you that this is not quite so. Without changing anything in the basic assumptions of Darwinism we can see that the behaviour of the individual, the way it makes use of its innate faculties, plays a relevant part, nay, plays the most relevant part in evolution. There is a very true kernel in Lamarck's view, namely that there is an irrescindable causal connection between the functioning, the actually being put to profitable use of a character - an organ, any property or ability or bodily feature - and its being developed in the course of generations, and gradually improved for the pur- poses for which it is profitably used. This connection, I say, between being used and being improved was a very correct cognition of Lamarck's, and it subsists in our present Darwinistic outlook, but it is easily overlooked on viewing Darwinism superficially. The course of events is almost the same as if Lamarckism were right, only the 'mechanism' by which things happen is more complicated than Lamarck thought. The point is not very easy to explain or to grasp, and so it may be useful to summarize the result in advance. To avoid vagueness, let us think of an organ, though the feature in question might be any property, habit, device, behaviour, or even any small addition to, or modification of, such a feature. Lamarck thought that the organ (a) is used, (b) is thus improved, and (c) the improvement is transmitted to the offspring. This is wrong. We have to think that the organ (a) undergoes chance variations, *(b)* the profitably used ones are accumulated or at least accentuated by selection, (c) this continues from generation to generation, the selected muta- tions constituting a lasting improvement. The most striking simulation of Lamarckism occurs - according to Julian Hux- ley - when the initial variations that inaugurate the process are not true mutations, not yet of the inheritable type. Yet, if profitable, they may be accentuated by what he calls organic selection, and, so to speak, pave the way for true mutations to be immediately seized upon when they happen to turn up in the 'desirable' direction.

Let us now go into some details. The most important point is to see that a new character, or modification of a character, acquired by variation, by mutation or by mutation plus some little selection, may easily arouse the organism in relation to its environment to an activity that tends to increase the usefulness of that character and hence the 'grip' of selection on it. By possessing the new or changed character the individual may be caused to change its environment - either by actually transforming it, or by migration - or it may be caused to change its behaviour towards its environment, all this in a fashion so as strongly to reinforce the usefulness of the new character and thus to speed up its further selective improve- ment in the same direction.

This assertion may strike you as daring, since it seems to require purpose on the side of the individual, and even a high degree of intelligence. But I wish to make the point that my statement, while it includes, of course, the intelligent, purposeful behaviour of the higher animals, is by no means restricted to them. Let us give a few examples:

Not all the individuals of a population have exactly the same environment. Some of the flowers of a wild species happen to grow in the shadow, some in sunny spots, some in the higher ranges of a lofty mountain-slope, some in the lower parts or in the valley. A mutation - say hairy foliage - which is beneficial at higher altitudes, will be favoured by selection in the higher ranges but will be 'lost' in the valley. The effect is the same as if the hairy mutants had migrated towards an environment that will favour further mutations that occur in the same direction.

Another example: their ability to fly enables birds to build their nests high up in the trees where their young ones are less accessible to some of their enemies. Primarily those who took to it had a selectional advantage. The second step is that this kind of abode was bound to select the proficient fliers among the young ones. Thus a certain ability to fly produces a change of environment, or behaviour towards the environment, which favours an accumulation of the same ability.

The most remarkable feature among living beings is that they are divided into species which are, many of them, so incredibly specialized on quite particular, often tricky per- formances, on which especially they rely for survival. A zoological garden is almost a curiosity show, and would be much more so, could it include an insight into the life-history of insects. Non-specialization is the exception. The rule is specialization in peculiar studied tricks which 'nobody would think ofif nature had not made them'. It is difficult to believe that they have all resulted from Darwinian 'accumulation by chance'. Whether one wants it or not, one is taken by the impression of forces or tendencies away from 'the plain and simple' in certain directions towards the complicated. The 'plain and simple' seems to represent an unstable state of affairs. A departure from it provokes forces - so it seems - towards a further departure, and in the same direction. That would be difficult to understand if the development of a particular device, mechanism, organ, useful behaviour, were produced by a long pearlstring of chance events, independent of each other, such as one is used to thinking of in terms of Darwin's original conception. Actually, I believe, only the first small start 'in a certain direction' has this structure. It produces itself circumstances which 'hammer the plastic material' - by selection - more and more systematically in the direction of the advantage gained at the outset. In metaphori- cal speech one might say: the species has found out in which direction its chance in life lies and pursues this path.

FEIGNED LAMARCKISM

We must try to understand in a general way, and to formulate in a non-animistic fashion, how a chance-mutation, which gives the individual a certain advantage and favours its survival in a given environment, should tend to do more than that, namely to increase the opportunities for its being profitably made use of, so as to concentrate on itself, as it were, the selective influence of the environment.

To reveal this mechanism let the environment be schem- atically described as an ensemble of favourable and unfavour- able circumstances. Among the first are food, drink, shelter, sunlight and many others, among the latter are the dangers from other living beings (enemies), poisons and the roughness of the elements. For brevity we shall refer to the first kind as 'needs' and to the second as 'foes'. Not every need can be obtained, not every foe avoided. But a living species must have acquired a behaviour that strikes a compromise in avoiding the deadliest foes and satisfying the most urgent needs from the sources of easiest access, so that it does survive. A favourable mutation makes certain sources more easily accessible, or reduces the danger from certain foes, or both. It thereby increases the chance of survival of the individuals endowed with it, but in addition it shifts the most favourable compromise, because it changes the relative weights of those needs or foes on which it bears. Individuals which - by chance or intelligence - change their behaviour accordingly will be more favoured, and thus selected. This change of behaviour is not transmitted to the next generation by the genom, not by direct inheritance, but this does not mean that it is not transmitted. The simplest, most primitive example is afforded' by our species of flowers (with a habitat along an extended mountain slope) that develops a hairy mutant. The hairy mutants, favoured mainly in the top ranges, disperse their seeds in such areas so that the next generation of 'hairies', taken as a whole, has 'climbed up the slope', as it were, 'to make better use of their favourable mutation'.

In all this one must bear in mind that as a rule the whole situation is extremely dynamic, the struggle is a very stiff one. In a fairly prolific population that, at the time, survives without appreciably increasing, the foes usually overpower the needs - individual survival is an exception. Moreover, foes and needs are frequently coupled, so that a pressing need can only be met by braving a certain foe. (For instance, the antelope has to come to the river for drink, but the lion knows the place just as well as he.) The total pattern of foes and needs is intricately interwoven. Thus a slight reduction of a certain danger by a given mutation may make a considerable difference for those mutants who brave that danger and thereby avoid others. This may result in a noticeable selection not only of the genetic feature in question but also with regard to the (intended or haphazard) skill in using it. That kind of behaviour is transmitted to the offspring by example, by learning, in a generalized sense of the word. The shift of behaviour, in turn, enhances the selective value of any further mutation in the same direction.

The effect of such a display may have great similarity with the mechanism as pictured by Lamarck. Though neither an acquired behaviour nor any physical changes that it entails are directly transmitted to the offspring, yet behaviour has an important say in the process. But the causal connection is not what Lamarck thought it to be, rather just the other way round. It is not that the behaviour changes the physique of the parents and, by physical inheritance, that of the offspring. It is the physical change in the parents that modifies - directly or indirectly, by selection - their behaviour; and this change of behaviour is, by example or teaching or even more primi- tively, transmitted to the progeny, along with the physical change carried by the genom. Nay, even if the physical change is not yet an inheritable one, the transmission of the induced behaviour 'by teaching' can be a highly efficient evolutionary factor, because it throws the door open to receive future inheritable mutations with a prepared readiness to make the best use of them and thus to subject them to intense selection.

GENETIC FIXATION OF HABITS AND SKILLS

Onemight object that what we have here described may happen occasionally, but cannot continue indefinitely to form the essential mechanism of adaptive evolution. For the change of behaviour itself is not transmitted by physical inheritance, by the hereditary substance, the chromosomes. At first, therefore, it is certainly not fixed genetically and it is difficult to see how it should ever come to be incorporated in the hereditary treasure. This is an important problem in itself. For we do know that habits are inherited as, for instance, habits of nestbuilding in the birds, the various habits of cleanliness we observe in our dogs and cats, to mention a few obvious examples. If this could not be understood along orthodox Darwinian lines, Darwinism would have to be abandoned. The question becomes of singular significance in its application to man, since we wish to infer that the striving and labouring of a man during his lifetime constitute an integrating contribution to the development of the species, in the quite proper biological sense. I believe the situation to be, briefly, as follows.

According to our assumptions the behaviour changes paral- lel those of the physique, first as a consequence of a chance change in the latter, but very soon directing the further selectional mechanism into definite channels, because, according as behaviour has availed itself of the first rudimentary benefits, only further mutations in the same direction have any selective value. But as (let me say) the new organ develops, behaviour becomes *more* and more bound up with its mere possession. Behaviour and physique merge into one. You simply cannot possess clever hands without using them for obtaining your aims, they would be in your way (as they often are to an amateur on the stage, because he has only fictitious aims). You cannot have efficient wings without attempting to fly. You cannot have a modulated organ of speech without trying to imitate the noises you hear around you. To distinguish between the possession of an organ and the urge to use it and to increase its skill by practice, to regard them as two different characteristics of the organism in question, would be an artificial distinction, made possible by an abstract language but having no counterpart in nature. We must, of course, not think that 'behaviour' after all gradually intrudes into the chromosome structure (or what not) and acquires 'loci' there. It is the new organs themselves (and they do become genetically fixed) that carry along with them the habit and the way of using them. Selection would be powerless in 'producing' a new organ if selection were not aided all along by the organism's making appropriate use of it. And this is very essential. For thus, the two things go quite parallel and are ultimately, or indeed at every stage, fixed genetically as one thing: *a used organ* - as if Lamarck were right.

It is illuminating to compare this natural process with the making of an instrument by man. At first sight there appears to be a marked contrast. If we manufacture a delicate mechanism, we should in most cases spoil it if we were impatient and tried to use it again and again long before it is finished. Nature, one is inclined to say, proceeds differently. She cannot produce a new organism and its organs otherwise than whilst they are continually used, probed, examined with regard to their efficiency. But actually this parallel is wrong. The making of a single instrument by man corresponds to ontogenesis, that is, to the growing up of a single individual from the seed to maturity. Here too interference is not welcome. The young ones must be protected, they must not be put to work before they have acquired the full strength and skill of their species. The true parallel of the evolutionary development of organisms could be illustrated, for example, by a historical exhibition of bicycles, showing how this machine gradually changed from year to year, from decade to decade, or, in the same way, of railway-engines, motor-cars, aeroplanes, typewriters, etc. Here, just as in the natural process, it is obviously essential that the machine in question should be continually used and thus improved; not literally improved by use, but by the experience gained and the alterations suggested. The bicycle, by the way, illustrates the case, mentioned before, of an old organism, which has reached the attainable perfection and has therefore pretty well ceased to undergo further changes. Still it is not about to become extinct!

DANGERS TO INTELLECTUAL EVOLUTION

Let us now return to the beginning of this chapter. We started from the question: is further biological development in man likely? Our discussion has, I believe, brought to the fore two relevant points.

The first is the biological importance of behaviour. By conforming to innate faculties as well as to the environment and by adapting itself to changes in either of these factors, behaviour, though not itself inherited, may yet speed up the process of evolution by orders of magnitude. While in plants and in the lower ranges of the animal kingdom adequate behaviour is brought about by the slow process of selection, in other words by trial and error, man's high intelligence enables him to enact it by choice. This incalculable advantage may easily outweigh his handicap of slow and comparatively scarce propagation, which is further reduced by the biologically dangerous regard not to let our offspring exceed the volume for which livelihood can be secured.

The second point, concerning the question whether biological development is still to be expected in man, is intimately connected with the first. In a way we get the full answer, namely, this will depend on us and our doing. We must not wait for things to come, believing that they are decided by irrescindable destiny. If we want it, we must do something about it. If not, not.Just as the political and social development and the sequence of historical events in general are not thrust upon us by the spinning of the Fates, but largely depend on our own doing, so our biological future, being nothing else but history on a large scale, must not be taken to be an unalterable destiny that is decided in advance by any Law of Nature. To us at any rate, who are the acting subjects in the play, it is not, even though to a superior being, watching us as we watch the birds and the ants, it might appear to be. The reason why man tends to regard history, in the narrower and in the wider sense, as a predestined happening, controlled by rules and laws that he cannot change, is very obvious. It is because every single individual feels that he by himself has very little say in the matter, unless he can put his opinions over to many others and persuade them to regulate their behaviour accordingly.

As regards the concrete behaviour necessary to secure our

biological future, I will only mention one general point that I consider of primary importance. We are, I believe, at the moment in grave danger of missing the 'path to perfection'. From all that has been said, selection is an indispensable requisite for biological development. If it is entirely ruled out, development stops, nay, it may be reversed. To put it in the words of Julian Huxley: ' ... the preponderance of degener- ative (loss) mutation will result in degeneration of an organ when it becomes useless and selection is accordingly no longer acting on it to keep it up to the mark.'

Now I believe that the increasing mechanization and 'stupidization' of most manufacturing processes involve the serious danger of a general degeneration of our organ of intelligence. The more the chances in life of the clever and of the unresponsive worker are equalled out by the repression of handicraft and the spreading of tedious and boring work on the assembly line, the more will a good brain, clever hands and a sharp eye become superfluous. Indeed the unintelligent man, who naturally finds it easier to submit to the boring toil, will be favoured; he is likely to find it easier to thrive, to settle down and to beget offspring. The result may easily amount even to a negative selection as regards talents and gifts.

The hardship of modern industrial life has led to certain institutions calculated to mitigate it, such as protection of the workers against exploitation and unemployment, and many other welfare and security measures. They are duly regarded as beneficial and they have become indispensable. Still we cannot shut our eyes to the fact that, by alleviating the responsibility of the individual to look after himself and by levelling the chances of every man, they also tend to rule out the competition of talents and thus to put an efficient brake on biological evolution. I realize that this particular point is highly controversial. One may make a strong case that the care for our present welfare must override the worry about our evolutionary future. But fortunately, so I believe, they go together according to my main argument. Next to want, boredom has become the worst scourge in our lives. Instead of letting the ingenious machinery we have invented produce an increasing amount of superfluous luxury, we must plan to develop it so that it takes off human beings all the unintelligent, mechanical, 'machine-like' handling. The machine must take over the toil for which man is too good, not man the work for which the machine is too expensive, as comes to pass quite often. This will not tend to make production cheaper, but those who are engaged in it happier. There is small hope of putting this through as long as the competition between big firms and concerns all over the world prevails. But this kind of competition is as uninteresting as it is biologically worthless. Our aim should be to reinstate in its place the interesting and intelligent competition of single human beings.

CHAPTER 3

The Principle of Objectivation

Nine years ago I put forward two general principles that form the basis of the scientific method, the principle of the under- standability of nature, and the principle of objectivation. Since then I have touched on this matter now and again, last time in my little book *Nature and the Greeks.*' I wish to deal here in detail with the second one, the objectivation. Before I say what I mean by that, let me remove a possible misunder- standing which might arise, as I came to realize from several reviews of that book, though I thought I had prevented it from the outset. It is simply this: some people seemed to think that my intention was to lay down the fundamental principles which *ought* to be at the basis of scientific method or at least which justly and rightly are at the basis of science and ought to be kept at all cost. Far from this, I only maintained and maintain that they *are* - and, by the way, as an inheritance from the ancient Greeks, from whom all our Wes tern science and scientific thought has originated.

The misunderstanding is not very astonishing. If you hear a scientist pronounce basic principles of science, stressing two of them as particularly fundamental and of old standing, it is natural to think that he is at least strongly in favour of them and wishes to impose them. But on the other hand, you see, science never imposes anything, science *states.* Science aims at nothing but making true and adequate statements about its object. The scientist only imposes two things, namely truth and sincerity, imposes them upon himself and upon other scientists. In the present case the object is science itself, as it has developed and has become and at present is, not as it *ought* to be or *ought* to develop in future.

Now let us turn to these two principles themselves. As regards the first, 'that nature can be understood', I will say here only a few words. The most astonishing thing about it is that it had to be invented, that it was at all necessary to invent it. It stems from the Milesian School, the *physiologoi.* Since then it has remained untouched, though perhaps not always uncontaminated. The present line in physics is possibly a quite serious contamination. The uncertainty principle, the alleged lack of strict causal connection in nature, may represent a step away from it, a partial abandonment. It would be interesting to discuss this, but I set my heart here on discussing the other principle, that which I called objectivation.

By this I mean the thing that is also frequently called the 'hypothesis of the real world' around us. I maintain that it amounts to a certain simplification which we adopt in order to master the infinitely intricate problem of nature. Without being aware of it and without being rigorously systematic about it, we exclude the Subject of Cognizance from the domain of nature that we endeavour to understand. We step with our own person back into the part of an onlooker who does not belong to the world, which by this very procedure becomes an objective world. This device is veiled by the following two circumstances. First, my own body (to which my mental activity is so very directly and intimately linked) forms part of the object (the real world around me) that I construct out of my sensations, perceptions and memories. Secondly, the bodies of other people form part of this objective world. Now I have very good reasons for believing that these other bodies are also linked up with, or are, as it were, the seats of spheres of consciousness. I can have no reasonable doubt about the existence or some kind of actualness of these foreign spheres of consciousness, yet I have absolutely no direct subjective access to any of them. Hence I am inclined to take them as something objective, as forming part of the real world around me. Moreover, since there is no distinction

between myself and others, but on the contrary full symmetry for all intents and purposes, I conclude that I myself also form part of this real material world around me. I so to speak put my own sentient self (which had constructed this world as a mental product) back into it - with the pandemonium of disastrous logical consequences that flow from the aforesaid chain of faulty conclusions. We shall point them out one by one; for the moment let me just mention the two most blatant antinomies due to our awareness of the fact that a moderately satisfying picture of the world has only been reached at the high price of taking ourselves out of the picture, stepping back into the role of a non-concerned observer.

The first of these antinomies is the astonishment at finding our world picture 'colourless, cold, mute'. Colour and sound, hot and cold are our immediate sensations; small wonder that they are lacking in a world model from which we have removed our own mental person.

The second is our fruitless quest for the place where mind acts on matter or vice-versa, so well known from Sir Charles Sherrington's honest search, magnificently expounded in *Man on his Nature.* The material world has only been constructed at the price of taking the self, that is, mind, out ofit, removing it; mind is not part of it; obviously, therefore, it can neither act on it nor be acted on by any of its parts. (This was stated in a very brief and clear sentence by Spinoza, see p. 1 22.)

I wish to go into more detail about some of the points I have made. First let me quote a passage from a paper ofC.G.Jung which has gratified me because it stresses the same point in quite a different context, albeit in a strongly vituperative fashion. While I continue to regard the removal of the Subject of Cognizance from the objective world picture as the high price paid for a fairly satisfactory picture, for the time being, Jung goes further and blames us for paying this ransom from an inextricably difficult situation. He says:

All science *(Wissenschaft)* however is a function of the soul, in which all knowledge is rooted. The soul is the greatest of all cosmic miracles, it is the *conditio sine qua non* of the world as an object. It is exceedingly astonishing that the Western world (apart from very rare exceptions) seems to have so little appreciation of this being so. The flood of external objects of cognizance has made the subject of all cognizance withdraw to the background, often to apparent non-existence.[1]

Of course Jung is quite right. It is also clear that he, being engaged in the science of psychology, is much more sensitive to the initial gambit in question, much more so than a physicist or a physiologist. Yet I would say that a rapid withdrawal from the position held for over 2,000 years is dangerous. We may lose everything without gaining more than some freedom in a special - though very important - domain. But here the problem is set. The relatively new science of psychology imperatively demands living-space, it makes it unavoidable to reconsider the initial gambit. This is a hard task, we shall not settle it here and now, we must be content at having pointed it out.

While here we found the psychologist Jung complaining about the exclusion of the mind, the neglect of the soul, as he terms it, in our world picture, I should now like to adduce in contrast, or perhaps rather as a supplement, some quotations of eminent representatives of the older and humbler sciences of physics and physiology, just stating the fact that 'the world of science' has become so horribly objective as to leave no room for the mind and its immediate sensations.

Some readers may remember A.S. Eddington's 'two writing desks'; one is the familiar old piece of furniture at which he is seated, resting his arms on it, the other is the scientific physical body which not only lacks all and every sensual qualities but in addition is riddled with holes; by far the greatest part of it is empty space, just nothingness, interspersed with innumerable tiny specks of something, the electrons and the nuclei whirling around, but always separated by distances at least roo,ooo times their own size. After having contrasted the two in his wonderfully plastic style he summarizes thus:

In the world of physics we watch a shadowgraph performance of familiar life. The shadow of my elbow rests on the shadow table as the shadow ink flows over the shadow paper ... The frank reali- zation that physical science is concerned with a world of shadows is one of the most significant of recent advances. [1]

Please note that the very recent advance does not lie in the world of physics itself having acquired this shadowy charac- ter; it had it ever since Democritus of Abdera and even before, but we were not aware of it; we thought we were dealing with the world itself; expressions like model or picture for the conceptual constructs of science came up in the second half of the nineteenth century, and not earlier, as far as I know.

Not much later Sir Charles Sherrington published his momentous *Man on his Nature.*[2] The book is pervaded by the honest search for objective evidence of the interaction between matter and mind. I stress the epithet 'honest', because it does need a very serious and sincere endeavour to look for some- thing which one is deeply convinced in advance cannot be found, because (in the teeth of popular belief) it does not exist. A brief summary of the result of this search is found on p. 35T

Mind, the anything perception can compass, goes therefore in our spatial world more ghostly than a ghost. Invisible, intangible, it is a thing not even of outline; it is not a 'thing'. It remains without sensual confirmation and remains without it forever.

In my own words I would express this by saying: Mind has erected the objective outside world of the natural philosopher out of its own stuff. Mind could not cope with this gigantic task otherwise than by the simplifying device of excluding itself - withdrawing from its conceptual creation. Hence the latter does not contain its creator.

I cannot convey the grandeur of Sherrington's immortal book by quoting sentences; one has to read it oneself. Still, I will mention a few of the more particularly characteristic.

Physical science ... faces us with the impasse that mind

per se

cannot play the piano - mind *per se* cannot move a finger of a hand

Then the impasse meets us. The blank of the 'how' of mind's leverage on matter. The inconsequence staggers us. Is it a misunder- standing? (p. 232).

Hold these conclusions drawn by an experimental physiol- ogist of the twentieth century against the simple statement of the greatest philosopher of the seventeenth century: B. Spinoza *(Ethics,* Pt m, Prop. 2):

Nee corpus mentem ad cogitandum, nee mens corpus ad motum, neque ad quietem, nee ad aliquid (si quid est) aliud determinare potest.

[Neither can the body determine the mind to think, nor the mind determine the body to motion or rest or anything else (if such there be).]

The impasse *is* an impasse. Are we thus not the doers of our deeds? Yet we feel responsible for them, we are punished or praised for them, as the case may be. It is a horrible antinomy. I maintain that it cannot be solved on the level of present-day science which is still entirely engulfed in the 'exclusion principle' - without knowing it - hence the antinomy. To realize this is valuable, but it does not solve the problem. You cannot remove the 'exclusion principle' by act of parliament as it were. Scientific attitude would have to be rebuilt, science must be made anew. Care is needed.

So we are faced with the following remarkable situation. While the stuff from which our world picture is built is yielded exclusively from the sense organs as organs of the mind, so that every man's world picture is and always remains a construct of his mind and cannot be proved to have any other existence, yet the conscious mind itself remains a stranger within that construct, it has no living space in it, you can spot it nowhere in space. We do not usually realize this fact, because we have entirely taken to thinking of the personality of a human being, or for that matter also that of an animal, as located in the interior of its body. To learn that it cannot really be found there is so amazing that it meets with doubt and hesitation, we are very loath to admit it. We have got used to localizing the conscious personality inside a person's head - I should say an inch or two behind the midpoint of the eyes. From there it gives us, as the case may be, understanding or loving or tender - or suspicious or angry looks. I wonder has it ever been noted that the eye is the only sense organ whose purely receptive character we fail to recognize in na'ive thought. Reversing the actual state of affairs, we are much more inclined to think of 'rays of vision', issuing from the eye, than of the 'rays of light' that hit the eyes from outside. You quite frequently find such a 'ray of vision' represented in a drawing in a comic paper, or even in some older schematic sketch intended to illustrate an optic instrument or law, a dotted line emerging from the eye and pointing to the object, the direction being indicated by an arrowhead at the far end. - Dear reader or, or better still, dear lady reader, recall the bright, joyful eyes with which your child beams upon you when you bring him a new toy, and then let the physicist tell you that in reality nothing emerges from these eyes; in reality their only objectively detectable function is, continually to be hit by and to receive light quanta. In reality! A strange reality! Something seems to be missing in it.

It is very difficult for us to take stock of the fact that the localization of the personality, of the conscious mind, inside the body is only symbolic, just an aid for practical use. Let us, with all the knowledge we have about it, follow such a 'tender look' inside the body. We do hit there on a supremely interesting bustle or, if you like, machinery. We find millions of cells of very specialized build in an arrangement that is unsurveyably intricate but quite obviously serves a very far-reaching and highly consummate mutual communication and collaboration; a ceaseless hammering of regular electro-chemical pulses which, however, change rapidly in their configuration, being conducted from nerve cell to nerve cell, tens of thousands of contacts being opened and blocked within every split second, chemical transformations being induced and maybe other changes as yet undiscovered. All this we meet and, as the science of physiology advances, we may trust that we shall come to know more and more about it. But now let us assume that in a particular case you eventually observe several efferent bundles of pulsating currents, which issue from the brain and through long cellular protrusions (motor nerve fibres), are conducted to certain muscles of the arm, which, as a consequence, tends a hesitating, trembling hand to bid you farewell - for a long, heart-rending separation; at the same time you may find that some other pulsating bundles produce a certain glandular secretion so as to veil the poor sad eye with a crape of tears. But nowhere along this way from the eye through the central organ to the arm muscles and the tear glands - nowhere, you may be sure, however far physiology advances, will you ever meet the personality, will you ever meet the dire pain, the bewildered worry within this soul, though their reality is to you so certain as though you suffered them yourself - as in actual fact you do! The picture that physiological analysis vouchsafes to us of any other human being, be it our most intimate friend, strikingly recalls to me Edgar Allan Poe's masterly story, which I am sure many a reader remembers well; I mean *The Masque of the Red Death.* A princeling and his retinue have withdrawn to an isolated castle to escape the pestilence of the red death that rages in the land. After a week or so of retirement they arrange a great dancing feast in fancy dress and mask. One of the masks, tall, entirely veiled, clad all in red and obviously intended to represent the pestilence allegorically, makes everybody shud- der, both for the wantonness of the choice and for the suspicion that it might be an intruder. At last a bold young man approaches the red mask and with a sudden jolt tears off veil and head-gear. It is found empty.

Now our skulls are not empty. But what we find there, in spite of the keen interest it arouses, is truly nothing when held against the life and the emotions of the soul.

To become aware of this may in the first moment upset one. To me it seems, on deeper thought, rather a consolation. If you have to face the body of a deceased friend whom you sorely miss, is it not soothing to realize that this body was never really the seat of his personality but only symbolically 'for practical reference'?

As an appendix to these considerations, those strongly interested in the physical sciences might wish to hear me pronounce on a line of ideas, concerning subject and object, that has been given great prominence by the prevailing school of thought in quantum physics, the protagonists being Niels Bohr, Werner Heisenberg, Max Born and others. Let me first give you a very brief description of their ideas. It runs as follows: [1]

We cannot make any factual statement about a given natural object (or physical system) without 'getting in touch' with it. This 'touch' is a real physical interaction. Even if it consists only in our 'looking at the object' the latter must be hit by light-rays and reflect them into the eye, or into some instrument of observation. This means that the object is affected by our observation. You cannot obtain any know- ledge about an object while leaving it strictly isolated. The theory goes on to assert that this disturbance is neither irrelevant nor completely surveyable. Thus after any number of painstaking observations the object is left in a state of which some features (the last observed) are known, but others (those interfered with by the last observation) are not known, or not accurately known. This state of affairs is offered as an explanation why no complete, gapless description of any physical object is ever possible.

If this has to be granted - and possibly it has to be granted -

then it flies in the face of the principle of understandability of nature. This in itself is no opprobrium. I told you at the outset that my two principles are not meant to be binding on science, that they only express what we had actually kept to in physical science for many, many centuries and what cannot easily be changed.

Personally I do not feel sure that our present knowledge as yet vindicates the change. I consider it possible that our models can be modified in such a fashion that they do not exhibit at any moment properties that cannot in principle be observed simultaneously - models poorer in simultaneous properties but richer in adaptability to changes in the environment. However, this is an internal question of

'*Sec* my *Srience and Humanism* (Cambridge Univcrsty Press, 19_<j1), p. +9 ·

physics, not to be decided here and now. But from the theory as explained before, from the unavoidable and unsurveyable interference of the measuring devices with the object under observation, lofty consequences of an epistemological nature have been drawn and brought to the fore, concerning the relation between subject and object. It is maintained that recent discoveries in physics have pushed forward to the mysterious boundary between the subject and the object. This boundary, so we are told, is not a sharp boundary at all. We are given to understand that we never observe an object without its being modified or tinged by our own activity in observing it. We are given to understand that under the impact of our refined methods of observation and of thinking about the results of our experiments that mysterious boundary between the subject and the object has broken down.

In order to criticize these contentions let me at first accept

the time-hallowed distinction or discrimination between object and subject, as many thinkers both in olden times have accepted it and in recent times still accept it. Among the philosophers who accepted it - from Democritus of Abdera down to the 'Old Man of Konigsberg' - there were few, if any who did not emphasize that all our sensations, perceptions and observations have a strong, personal, subjective tinge and do not convey the nature of the 'thing-in-itself, to use Kant's term. While some of these thinkers might have in mind only a more or less strong or slight distortion, Kant landed us with a complete resignation: never to know anything at all about his 'thing-in-itself. Thus the idea of subjectivity in all appearance is very old and familiar. What is new in the present setting is this: that not only would the impressions we get from our environment largely depend on the nature and the contingent state of our sensorium, but inversely the very environment that we wish to take in is modified by us, notably by the devices we set up in order to observe it.

Maybe this is so - to some extent it certainly is. May be that from the newly discovered laws of quantum physics this modification cannot be reduced below certain well-ascertained limits. Still I would not like to call this a direct influence of the subject on the object. For the subject, if anything, is the thing that senses and thinks. Sensations and thoughts do not belong to the 'world of energy', they cannot produce any change in this world of energy as we know from Spinoza and Sir Charles Sherrington.

All this was said from the point of view that we accept the time-hallowed discrimination between subject and object. Though we have to accept it in everyday life 'for practical reference', we ought, so I believe, to abandon it in philosophi- cal thought. Its rigid logical consequence has been revealed by Kant: the sublime, but empty, idea of the 'thing-in-itself about which we forever know nothing.

It is the same elements that go to compose my mind and the world. This situation is the same for every mind and its world, in spite of the unfathomable abundance of 'cross-references' between them. The world is given to me only once, not one existing and one perceived. Subject and object are only one. The barrier between them cannot be said to have broken down as a result of recent experience in the physical sciences, for this barrier does not exist.

CHAPTER 4
The Arithmetical Paradox: The Oneness of Mind

The reason why our sentient, percipient and thinking ego is met nowhere within our scientific world picture can easily be indicated in seven words: because it is itself that world picture. It is identical with the whole and therefore cannot be contained in it as a part of it. But, of course, here we knock against the arithmetical paradox; there appears to be a great multitude of these conscious egos, the world however is only one. This comes from the fashion in which the world-concept produces itself. The several domains of 'private' conscious- nesses partly overlap. The region common to all where they all overlap is the construct of the 'real world around us'. With all that an uncomfortable feeling remains, prompting such questions as: Is my world really the same as yours? Is there *one* real world to be distinguished from its pictures introjected by way of perception into every one of us? And if so, are these pictures like unto the real world or is the latter, the world 'in itseir, perhaps very diflerent from the one we perceive?

Such questions are ingenious, but in my opinion very apt to confuse the issue. They have no adequate answers. They all are, or lead to, antinomies springing from the one source, which I called the arithmetical paradox; the *many* conscious egos from whose mental experiences the *one* world is concoc- ted. The solution of this paradox of numbers would do away with all the questions of the aforesaid kind and reveal them, I dare say, as sham questions.

There are two ways out of the number paradox, both appearing rather lunatic from the point of view of present scientific thought (based on ancient Greek thought and thus thoroughly 'Western'). One way out is the multiplication of the world in Leibniz's fearful doctrine of monads: every monad to be a world by itself, no communication between them; the monad 'has no windows', it is 'incommunicado'. That none the less they all agree with each other is called 'pre-established harmony'. I think there are few to whom this suggestion appeals, nay who would consider it as a mitigation at all of the numerical antinomy.

There is obviously only one alternative, namely the unifica- tion of minds or consciousnesses. Their multiplicity is only apparent, in truth there is only one mind. This is the doctrine of the Upanishads. And not only of the Upanishads. The mystically experienced union with God regularly entails this attitude unless it is opposed by strong existing prejudices; and this means that it is less easily accepted in the West than in the East. Let me quote as an example outside the Upanishads an Islamic Persian mystic of the thirteenth century, Aziz Nasafi. I am taking it from a paper by Fritz Meyer[1] and translating from his German translation:

> On the death of any living creature the spirit returns to the spiritual world, the body to the bodily world. In this however only the bodies are subject to change. The spiritual world is one single spirit who stands like unto a light behind the bodily world and who, when any single creature comes into being, shines through it as through a window. According to the kind and size of the window less or more light enters the world. The light itself however remains unchanged.

Ten years ago Aldous Huxley published a precious volume which he called *The Perennial Philosophy*[2] and which is an anthology from the mystics of the most various periods and the most various peoples. Open it where you will and you find many beautiful utterances of a similar kind. You are struck by the miraculous agreement between humans of different race, different religion, knowing nothing about each other's exist- ence, separated by centuries and millennia, and by the greatest distances that there are on our globe.

Still, it must be said that to Western thought this

doctrine has little appeal, it is unpalatable, it is dubbed fantastic, unscientific. Well, so it is because our science - Greek science - is based on objectivation, whereby it has cut itself off from an adequate understanding of the Subject of Cognizance, of the mind. But I do believe that this is precisely the point where our present way of thinking does need to be amended, perhaps by a bit of blood-transfusion from Eastern thought. That will not be easy, we must beware of blunders - blood-transfusion always needs great precaution to prevent clotting. We do not wish to lose the logical precision that our scientific thought has reached, and that is unparalleled any- where at any epoch.

Still, one thing can be claimed in favour of the mystical teaching of the 'identity' of all minds with each other and with the supreme mind - as against the fearful monadology of Leibniz. The doctrine of identity can claim that it is clinched by the empirical fact that consciousness is never experienced in the plural, only in the singular. Not only has none of us ever experienced more than one consciousness, but there is also no trace of circumstantial evidence of this ever happening any- where in the world. If I say that there cannot be more than one consciousness in the same mind, this seems a blunt tautology- we are quite unable to imagine the contrary.

Yet there are cases or situations where we would expect and nearly require this unimaginable thing to happen, if it can happen at all. This is the point that I should like to discuss now in some detail, and to clinch it by quotations from Sir Charles Sherrington, who was at the same time (rare event!) a man of highest genius and a sober scientist. For all I know he had no bias towards the philosophy of the Upanishads. My purpose in this discussion is to contribute perhaps to clearing the way for a future assimilation of the doctrine of identity with our own scientific world view, without having to pay for it by a loss of soberness and logical precision.

I said just now that we are not able even to imagine a plurality of consciousnesses in one mind. We can pronounce these words all right, but they are not the description of any thinkable experience. Even in the pathological cases of a 'split personality' the two persons alternate, they never hold the field jointly; nay this is just the characteristic feature, that they know nothing about each other.

When in the puppet-show of dream we hold in hand the strings of quite a number of actors, controlling their actions and their speech, we are not aware of this being so. Only one of them is myself, the dreamer. In him I act and speak immediately, while I may be awaiting eagerly and anxiously what another one will reply, whether he is going to fulfil my urgent request. That I could really let him do and say whatever I please does not occur to me - in fact it is not quite the case. For in a dream of this kind the 'other one' is, I dare say, mostly the impersonation of some serious obstacle that opposes me in waking life and of which I have actually no control. The strange state of affairs, described here, is quite obviously the reason why most people of old firmly believed that they were truly in communication with the persons, alive or deceased, or, maybe, gods or heroes, whom they met in their dreams. It is a superstition that dies hard. On the verge of the sixth century B.C. Heraclitus of Ephesus definitely pronounced against it, with a clarity not often met with in his sometimes very obscure fragments. But Lucretius Carus, who believed himself to be the protagonist of enlightened thought, still holds on to this superstition in the first century B.c. In our days it is probably rare, but I doubt that it is entirely extinct.

Let me turn to something quite different. I find it utterly impossible to form an idea about either how, for example, my own conscious mind (that I feel to be *one)* should have originated by integration of the consciousnesses of the cells (or some of them) that form my body, or how it should at every moment of my life be, as it were, their resultant. One would think that such a 'commonwealth of cells' as each of us is would be the occasion *par excellence* for mind to exhibit plurality if it were at all able to do so. The expression 'commonwealth' or 'state of cells' *(Zellstaat)* is nowadays no longer to be regarded as a metaphor. Listen to Sherrington:

Todeclare that, of the component cells that go to make us up, each *one* is an individual self-centred life is no *mere* phrase. It is not a *mere* convenience for descriptive purposes. The cell as a component of the body is not only a visibly demarcated unit but a unit-life *centred* on itself. It leads its own life ... The cell is a unit-life, and our life which in its turn is a unitary life consists utterly of the cell-lives.[1]

But this story can be followed up in more detail and more concretely. Both the pathology of the brain and physiological investigations on sense perception speak unequivocally in favour of a regional separation of the sensorium into domains whose far-reaching independence is amazing because it would let us expect to find these regions associated with independent domains of the mind; but they are not. A particularly characteristic instance is the following. If you look at a distant landscape first in the ordinary way with both eyes open, then with the right eye alone, shutting the left, then the other way round, you find no noticeable difference. The psychic visional space is in all three cases identically the same. Now this might very well be due to the fact that from corresponding nerve-ends on the retina the stimulus is transferred to the same centre in the brain where 'the perception is manufactured' - just as, for example, in my house the knob at the entrance door and the one in my wife's bedroom activate the same bell, situated above the kitchen door. This would be the easiest explanation; but it is wrong.

Sherrington tells us of very interesting experiments on the thresnold frequency of flickering. I shall try to give you as brief an account as possible. Think of a miniature lighthouse set up in the laboratory and giving off a great many flashes per second, say 40 or 60 or 80 or 100. As you increase the frequency of the flashes the flickering disappears at a definite frequency, depending on the experimental details; and the onlooker, whom we suppose to watch with both eyes in the ordinary way, sees then a continuous light.[2] Let this threshold frequency be 60 per second in given circumstances. Now in a second experiment, with nothing else changed, a suitable contraption allows only every second flash to reach the right eye, every other flash to reach the left eye, so that every eye receives only 30 flashes per second. If the stimuli were conducted to the same physiological centre, this should make no difference: if I press the button before my entrance door, say every two seconds, and my wife does the same in her bedroom, but alternately with me, the kitchen bell will ring every second, just the same as if one of us had pressed his button every second or both of us had done so synchronously every second. How- ever, in the second flicker experiment this is not so. Thirty flashes to the right eye plus alternating 30 flashes to the left are far from sufficient to remove the sensation of flickering; double the frequency is required for that, namely, 60 to the right and 60 to the left, if both eyes are open. Let me give you the main conclusion in Sherrington's own words:

It is not spatial conjunction of cerebral mechanism which combines the two reports ... It is much as though the right- and left-eye images were seen each by one of two observers and the minds of the two observers were combined to a single mind. It is as though the right-eye and left-eye perceptions arc elaborated singly and then psychically combined to one ... It is as if each eye had a separate sensorium of considerable dignity proper to itself, in which mental processes based on that eye were developed up to even full percep- tual levels. Such would amount physiologically to a visual sub- brain. There would be two such sub--brains, one for the right eye and one for the left eye. Contemporaneity of action rather than structural union seems to provide their mental collaboration.[1]

This is followed by very general considerations, of which I shall again pick out only the most characteristic passages:

Are there thus quasi-independent sub-brains based on the several modalities of sense? In the roof brain the old 'five' senses instead of being merged inextricably in one another and further submerged under mechanism of higher order are still plain to find, each demarcated in its separate sphere. How far is the mind a collection of quasi-independent perceptual minds integrated psychically in large measure by temporal concurrence of experience? ... When it is a question of 'mind' the nervous system does not integrate itself by centralization upon a pontifical cell. Rather it elaborates a millionfold democracy whose each unit is a cell ... the concrete life compounded of sublives reveals, although integrated, its additive nature and declares itself an affair of minute foci of life acting together ... When however we turn to the mind there is nothing of all this. The single nerve-cell is never a miniature brain. The cellular constitution of the body need not be for any hint of it from 'mind' ... A single pontifical brain-cell could not assure to the mental reaction a character more unified, and non-atomic than does the roof-brain's multitudinous sheet of cells. Matter and energy seem granular in structure, and so does 'life', but not so mind.

I have quoted you the passages which have most impressed me. Sherrington, with his superior knowledge of what is actually going on in a living body, is seen struggling with a paradox which in his candidness and absolute intellectual sincerity he does not try to hide away or explain away (as many others would have done, nay have done), but he almost brutally exposes it, knowing very well that this is the only way of driving any problem in science or philosophy nearer towards its solution, while by plastering it over with 'nice' phrases you prevent progress and make the antinomy perennial ('not forever, but until someone notices your fraud). Sherrington's paradox too is an arithmetical paradox, a paradox of numbers, and it has, so I believe, very much to do with the one to which I had given this name earlier in this chapter, though it is by no means identical with it. The previous one was, briefly, the *one* world crystallizing out of the many minds. Sherrington's is the *one* mind, based ostensibly on the many cell-lives or, in another way, on the manifold sub-brains, each of which seems to have such a considerable dignity proper to itself that we feel impelled to associate a sub-mind with it. Yet we know that a sub-mind is an atrocious monstrosity, just as is a plural-mind - neither having any counterpart in anybody's experience, neither being in any way imaginable.

I submit that both paradoxes will be solved (I do not pretend to solve them here and now) by assimilating into our Western build of science the Eastern doctrine of identity.

Mind is by its very nature a *singulare tantum.* I should say: the over-all number of minds is just one. I venture to call it indestructible since it has a peculiar timetable, namely mind is always *now.* There is really no before and after for mind. There is only a now that includes memories and expectations. But I grant that our language is not adequate to express this, and I also grant, should anyone wish to state it, that I am now talking religion, not science - a religion, however, not opposed to science, but supported by what disinterested scientific research has brought to the fore.

Sherrington says: 'Man's mind is a recent product of our planet's side."

I agree, naturally. If the first word (man's) were left out, I would not. We dealt with this earlier, in chapter 1. It would seem queer, not to say ridiculous, to think that the contem- plating, conscious mind that alone reflects the becoming of the world should have made its appearance only at some time in the course of this 'becoming', should have appeared contin- gently, associated with a very special biological contraption which in itself quite obviously discharges the task of facilitat- ing certain forms of life in maintaining themselves, thus favouring their preservation and propagation: forms of life that were late-comers and have been preceded by many others that maintained themselves without that particular contrap- tion (a brain). Only a small fraction of them (if you count by species) have embarked on 'getting themselves a brain'. And before that happened, should it all have been a performance to empty stalls? Nay, may we call a world that nobody contem- plates even that? When an archaeologist reconstructs a city or a culture long bygone, he is interested in human life in the past, in actions, sensations, thoughts, feelings, in joy and sorrow of humans, displayed there and then. But a world existing for many millions of years without any mind being aware of it, contemplating it, is it anything at all? Has it existed? For do not let us forget: to say, as we did, that the becoming of the world is reflected in a conscious mind is but a cliche, a phrase, a metaphor that has become familiar to us. The world is given but once. Nothing is reflected. The original and the mirror-image are identical. The world extended in space and time is but our representation (*Vorstellung).* Experi- ence does not give us the slightest clue of its being anything besides that- asBerkeley was well aware.

But the romance of a world that had existed for many millions of years before it, quite contingently, produced brains in which to look at itself has an almost tragic continuation that I should like to describe again in Sherrington's words:

The universe of energy is we are told running down. It tends fatally towards an equilibrium which shall be final. An equilibrium in which life cannot exist. Yet life is being evolved without pause. Our planet in its surround has evolved it and is evolving it. And with it evolves mind. If mind is not an energy-system how will the running down of the universe affect it? Can it go unscathed? Always so far as we know the finite mind is attached to a running energy-system. When that energy-system ceases to run what of the mind which runs with it? Will the universe which elaborated and is elaborating the finite mind then let it perish?'

Such considerations are in some way disconcerting. The thing that bewilders us is the curious double role that the conscious mind acquires. On the one hand it is the stage, and the only stage on which this whole world-process takes place, or the vessel or container that contains it all and outside which there is nothing. On the other hand we gather the impression, maybe the deceptive impression, that within this world-bustle the conscious mind is tied up with certain very particular organs (brains), which while doubtless the most interesting contraption in animal and plant physiology are yet not unique, not *sui generis;* for like so many others they serve after all only to maintain the lives of their owners, and it is only to this that they owe their having been elaborated in the process of speciation by natural selection.

Sometimes a painter introduces into his large picture, or a poet into his long poem, an unpretending subordinate charac- ter who is himself. Thus the poet of the *Odyssey* has, I suppose, meant himself by the blind bard who in the hall of the Phaea- cians sings about the battles of Troy and moves the battered hero to tears. In the same way we meet in the song of the Nibelungs, when they traverse the Austrian lands, with a poet who is suspected to be the author of the whole epic. In Diirer's *All-Saints* picture two circles of believers are gathered in prayer around the Trinity high up in the skies, a circle of the blessed above, and a circle of humans on the earth. Among the latter are kings and emperors and popes, but also, if I am not mistaken, the portrait of the artist himself, as a humble side- figure that might as well be missing.

To me this seems to be the best simile of the bewildering

double role of mind. On the one hand mind is the artist who has produced the whole; in the accomplished work, however, it is but an insignificant accessory that might be absent without detracting from the total effect.

Speaking without metaphor we have to declare that we are here faced with one of these typical antinomies caused by the fact that we have not yet succeeded in elaborating a fairly understandable outlook on the world without retiring our own mind, the producer of the world picture, from it, so that mind has no place in it. The attempt to press it into it, after all, necessarily produces some absurdities.

Earlier I have commented on the fact that for this same reason the physical world picture lacks all the sensual qualities that go to make up the Subject of Cognizance. The model is colourless and soundless and unpalpable. In the same way and for the same reason the world of science lacks, or is deprived of, everything that has a meaning only in relation to the con- sciously contemplating, perceiving and feeling subject. I mean in the first place the ethical and aesthetical values, any values of any kind, everything related to the meaning and scope of the whole display. All this is not only absent but it cannot, from the purely scientific point of view, be inserted organically. If one tries to put it in or on, as a child puts colour on his uncoloured painting copies, it will not fit. For anything that is made to enter this world model willy-nilly takes the form of scientific assertion of facts; and as such it becomes wrong.

Life is valuable in itself. 'Be reverent towards life' is how Albert Schweitzer has framed the fundamental commandment of ethics. Nature has no reverence towards life. Nature treats life as though it were the most valueless thing in the world. Produced million-fold it is for the greatest part rapidly annihilated or cast as prey before other life to feed it. This precisely is the master-method of producing ever-new forms of life. 'Thou shalt not torture, thou shalt not inflict pain!' Nature is ignorant of this commandment. Its creatures depend upon racking each other in everlasting strife.

'There is nothing either good or bad but thinking makes it so.' No natural happening is in itself either good or bad, nor is it in itself either beautiful or ugly. The values are missing, and quite particularly meaning and end are missing. Nature does not act by purposes. If in German we speak of a purposeful *(zweckmassig)* adaptation of an organism to its environment, we know this to be only a convenient way of speech. Ifwe take it literally, we are mistaken. We are mistaken within the frame of our world picture. In it there is only causal linkage.

Most painful is the absolute silence of all our scientific investigations towards our questions concerning the meaning and scope of the whole display. The more attentively we watch it, the more aimless and foolish it appears to be. The show that is going on obviously acquires a meaning only with regard to the mind that contemplates it. But what science tells us about this relationship is patently absurd: as if mind had only been produced by that very display that it is now watching and would pass away with it when the sun finally cools down and the earth has been turned into a desert of ice and snow.

Let me briefly mention the notorious atheism of science which comes, of course, under the same heading. Science has to suffer this reproach again and again, but unjustly so. No personal god can form part of a world model that has only become accessible at the cost of removing everything personal from it. We know, when God is experienced, this is an event as real as an immediate sense perception or as one's own personality. Like them he must be missing in the space-time picture. I do not find God anywhere in space and time - that is what the honest naturalist tells you. For this he incurs blame from him in whose catechism is written: God is spirit.

CHAPTER 5
Science and Religion

Can science vouchsafe information on matters ofreligion? Can the results of scientific research be of any help in gaining a reasonable and satisfactory attitude towards those burning questions which assail everyone at times? Some of us, in particular healthy and happy youth, succeed in shoving them aside for long periods; others, in advanced age, have satisfied themselves that there is no answer and have resigned them- selves to giving up looking for one, while others again are haunted throughout their lives by this incongruity of our intellect, haunted also by serious fears raised by time- honoured popular superstition. I mean mainly the questions concerned with the 'other world', with 'life after death', and all that is connected with them. Notice please that I shall not, of course, attempt to answer *these* questions, but only the much more modest one, whether science can give any information about them or aid our - to many ofus unavoidable - thinking about them.

To begin with, in a very primitive way it certainly can, and has done so without much ado. I remember seeing old prints, geographical maps of the world, so I believe, including hell, purgatory and heaven, the former being placed deep under- ground, the latter high above in the skies. Such representa- tions were not meant purely allegorically (as they might be in later periods, for example, in Diirer's famous *All-Saints* pic- ture); they testify to a crude belief quite popular at the time. Today no church requests the faithful to interpret its dogmas in this materialistic fashion, nay it would seriously discourage such an attitude. This advancement has certainly been aided by our knowledge of the interior of our planet (scanty though it be), of the nature of volcanoes, of the composition of our atmosphere, of the probable history of the solar system and of the structure of the galaxy and the universe. No cultured person would expect to find these dogmatic figments in any region of that part of space which is accessible to our investigation, I daresay not even in a region continuing that space but inaccessible to research; he would give them, even if convinced of their reality, a spiritual standing. I will not say that with deeply religious persons such enlightenment had to await the aforesaid findings of science, but they have certainly helped in eradicating materialistic superstition in those matters.

However, this refers to a rather primitive state of mind. There are points of greater interest. The most important contributions from science to overcome the baffling questions 'Who are we really? Where have I come from and where am I going?' - or at least to set our minds at rest - I say, the most appreciable help science has offered us in this is, in my view, the gradual idealization of time. In thinking of this the names of three men obtrude themselves upon us, though many others, including non-scientists, have hit on the same groove, such as St Augustine of Hippo and Boethius; the three are Plato, Kant and Einstein.

The first two were not scientists, but their keen devotion to

philosophic questions, their absorbing interest in the world, originated from science. In Plato's case it came from math- ematics and geometry (the 'and' would be out of place today, but not, I think, in his time). What has endowed Plato's life-work with such unsurpassed distinction that it shines in undiminished splendour after more than two thousand years? For all we can tell, no special discovery about numbers or geometrical figures is to his credit. His insight into the material world of physics and life is occasionally fantastic and altogether inferior to that of others (the sages from Thales to Democritus) who lived, some of them more than a century, before his time; in knowledge of nature he was widely surpassed by his pupil Aristotle and by Theophrastus. To all but his ardent worshippers long passages in his dialogues give the impression of a gratuitous quibbling on words, with no desire to define the meaning of a word, rather in the belief that the word itself will display its content if you turn it round and round long enough. His social and political Utopia, which failed and put him into grave danger when he tried to promote it practically, finds few admirers in our days, that have sadly experienced the like. So what made his fame?

In my opinion it was this, that he was the first to envisage the idea of timeless existence and to emphasize it - against reason - as a reality, more real than our actual experience; this, he said, is but a shadow of the former, from which all experienced reality is borrowed. I am speaking of the theory of forms (or ideas). How did it originate? There is no doubt that it was aroused by his becoming acquainted with the teaching of Parmenides and the Eleatics. But it is equally obvious that this met in Plato with an alive congenial vein, an occurrence very much on the line of Plato's own beautiful simile that learning by reason has the nature of remembering knowledge, previously possessed but at the time latent, rather than that of discovering entirely new verities. However, Parmenides' ever- lasting, ubiquitous and changeless One has in Plato's mind turned into a much more powerful thought, the Realm of Ideas, which appeals to the imagination, though, of necessity, it remains a mystery. But this thought sprang, as I believe, from a very real experience, namely, that he was struck with admiration and awe by the revelations in the realm of numbers and geometrical figures - as many a man was after him and the Pythagoreans were before. **He** recognized and absorbed deeply into his mind the nature of these revelations, that they unfold themselves by pure logical reasoning, which makes us acquainted with true relations whose truth is not only unassailable, but is obviously there, forever; the relations held and will hold irrespective of our inquiry into them. A mathematical truth is timeless, it does not come into being when we discover it. Yet its discovery is a very real event, it may be an emotion like a great gift from a fairy.

The three heights of a triangle *(ABC)* meet at one point (0).

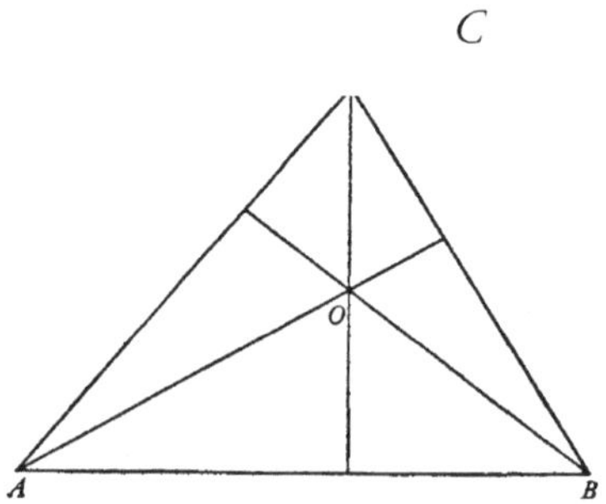

Fig. 1.

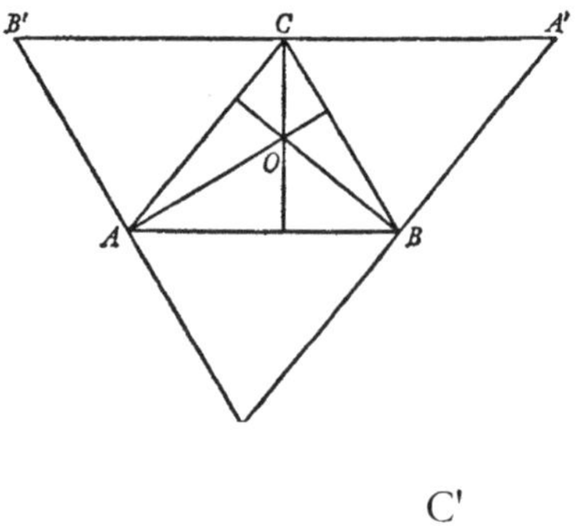

Fig. 2.

(Height is the perpendicular, dropped from a corner onto the side opposite to it, or onto its prolongation.) At first sight one does not see why they should; *any* three lines do not, they usually form a triangle. Now draw through every corner the parallel to the opposite side, to form the bigger triangle $A'B'C'$. It consists of four congruent triangles. The three heights of ABC are in the bigger triangle the perpendiculars erected in the middle of its sides, their 'symmetry lines'. Now the one erected at *C* must contain all the points that have the same distance from *A'* as from *B'*; the one erected at *B* contains all those points that have the same distance from *A'* as from C'. The point where these two perpendiculars meet has therefore the same distance from all three corners *A'*, *B'*, C', and must therefore lie also on the perpendicular erected at *A* because this one contains all points that have the same distance from *B'* as from C'. Q.E.D.

Every integer, except rand 2, is 'in the middle' of two prime numbers, or is their arithmetical mean; for instance

$8 = \frac{1}{2}(5 + 11) = \frac{1}{2}(3 + 13)$
$17 = \frac{1}{2}(3 + 31) = \frac{1}{2}(29 + 5) = \frac{1}{2}(23 + 11)$
$20 = \frac{1}{2}(11 + 29) = \frac{1}{2}(3 + 37)$.

As you see, there is usually more than one solution. The theorem is called Goldbach's and is thought to be true, though it has not been proved.

By adding the consecutive odd numbers, thus first taking just 1, then 1 + 3 = 4, then r + 3 + 5 = 9, then 1 + 3 + 5 + 7 = 16, you always get a square number, indeed you get in this way all square numbers, always the square of the number of odd numbers you have added. To grasp the genera- lity of this relation one may replace in the sum the summands of every pair that is equidistant from the middle (thus: the first and the last, then the first but one and the last but one, etc.) by their arithmetic mean, which is obviously just equal to the number of summands; thus, in the last of the above examples:

$4 + 4 + 4 + 4 = 4 \times 4$.

Let us now turn to Kant. It has become a commonplace that he taught the ideality of space and time and that this was a fundamental, if not the most fundamental part of his teaching. Like most ofit, it can be neither verified nor falsified, but it does not lose interest on this account (rather it gains; ifit could be proved or disproved it would be trivial). The meaning is that, to be spread out in space and to happen in a well-defined temporal order of 'before and after' is not a quality of the world that we perceive, but pertains to the perceiving mind which, in its present situation anyhow, cannot help registering anything that is offered to it according to these two card-indexes, space and time. It does not mean that the mind comprehends these order-schemes irrespective of, and before, any experience, but that it cannot help developing them and applying them to experience when this comes along, and particularly that this fact does not prove or suggest space and time to be an order-scheme inherent in that 'thing-in-itself which, as some believe, causes our experience.

It is not difficult to make a case that this is hum bug. No single man can make a distinction between the realm of his perceptions and the realm of things that cause it since, however detailed the knowledge he may have acquired about the whole story, the story is occurring only once not twice. The duplication is an allegory, suggested mainly by commu- nication with other human beings and even with animals; which shows that their perceptions in the same situation seem to be very similar to his own apart from insignificant differ- ences in the point of view - in the literal meaning of 'point of projection'. But even supposing that this compels us to consider an objectively existing world the cause of our percep- tions, as most people do, how on earth shall we decide that a common feature of all our experience is due to the constitution of our mind rather than a quality shared by all those objectively existing things? Admittedly our sense perceptions constitute our sole knowledge about things. This objective world remains a hypothesis, however natural. If we do adopt it, is it not by far the most natural thing to ascribe to that external world, and not to ourselves, all the characteristics that our sense perceptions find in it?

However, the supreme importance of Kant's statement does not consist in justly distributing the roles of the mind and its object - the world - between them in the process of 'mind forming an idea of the world', because, as I just pointed out, it is hardly possible to discriminate the two. The great thing was to form the idea that this *one thing* - mind or world - may well be capable of other forms of appearance that we cannot grasp and that do not imply the notions of space and time. This means an imposing liberation from our inveterate prejudice. There probably are other orders of appearance than the space-time-like. It was, so I believe, Schopenhauer who first read this from Kant. This liberation opens the way to belief, in the religious sense, without running all the time against the clear results which experience about the world as we know it and plain thought unmistakably pronounce. For instance - to speak of the most momentous example - experience as we know it unmistakably obtrudes the conviction that it cannot survive the destruction of the body, with whose life, as we know life, it is inseparably bound up. So is there to be nothing after this life? No. Not in the way of experience as we know it necessarily to take place in space and time. But, in an order of appearance in which time plays no part, this notion of 'after' is meaningless. Pure thinking cannot, of course, procure us a guarantee that there *is* that sort of thing, But it can remove the apparent obstacles to conceiving it as possible. That is what Kant has done by his analysis, and that, to my mind, is his philosophical importance.

I now come to speak about Einstein in the same context.

Kant's attitude towards science was incredibly nai:ve, as you will agree if you turn the leaves of his *Metaphysical Foundations of Science (Metaphysische Anfangsgrunde der Naturwissenschaft).* He accepted physical science in the form it had reached during his lifetime (1724-1804) as something more or less final and he busied himself to account for its statements philosophi- cally. This happening to a great genius ought to be a warning to philosophers ever after. He would show plainly that space was necessarily infinite and believed firmly that it was in the nature of the human mind to endow it with the geometrical properties summarized by

Euclid. In this Euclidean space a mollusc of matter moved, that is, changed its configuration as time went on. To Kant, as to any physicist of his period, space and time were two entirely different conceptions, so he had no qualms in calling the former the form of our external intuition, and time the form of our internal intuition *(Anschauung).* The recognition that Euclid's infinite space is not a necessary way of looking at the world of our experience and that space and time are better looked upon as one continuum of four dimensions seemed to shatter Kant's foundation - but actually did no harm to the more valuable part of his philosophy.

This recognition was left to Einstein (and several others, H. A. Lorentz, Poincare, Minkowski, for example). The mighty impact of their discoveries on philosophers, men-in- the-street, and ladies in the drawing-room is due to the fact that they brought it to the fore: even in the domain of our experience the spatio-temporal relations are much more intri- cate than Kant dreamed them to be, following in this all previous physicists, men-in-the-street and ladies in the drawing-room.

The new view has its strongest impact on the previous notion of time. Time is the notion of 'before and after'. The new attitude springs from the following two roots:

(i) The notion of 'before and after' resides on the 'cause and effect' relation. We know, or at least we have formed the idea, that one event *A* can cause, or at least modify, another event ***B,*** so that if *A* were not, then ***B*** were not, at least not in this modified form. For instance when a shell explodes, it kills a man who was sitting on it; moreover the explosion is heard at distant places. The killing may be simultaneous to the explosion, the hearing of the sound at a distant place will be later; but certainly none of the effects can be earlier. This is a basic notion, indeed it is the one by which also in everyday life the question is decided which of two events was later or at least not earlier. The distinction rests entirely on the idea that the effect cannot precede the cause. If we have reasons to think that *B* has been caused by *A,* or that it at least shows vestiges of *A,* or even if (from some circumstantial evidence) it is conceivable that it shows vestiges, then *B* is deemed to be certainly not earlier than *A.*

(2) Keep this in mind. The second root is the experimental and observational evidence that effects do not spread with arbitrarily high velocity. There is an upper limit, which incidentally is the velocity of light in empty space. In human measure it is very high, it would go round the equator about

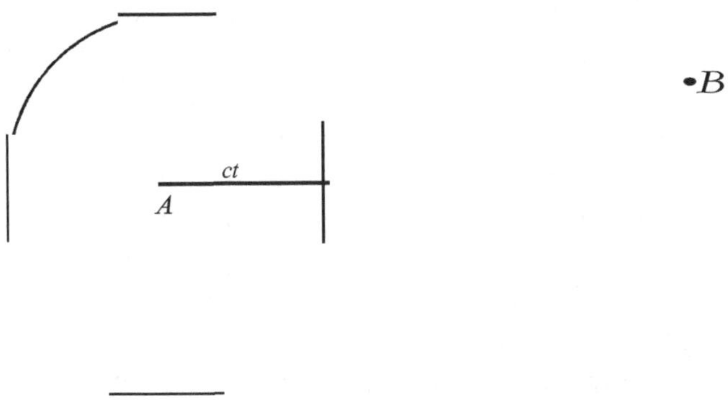

Fig. 3.

seven times in one second. Very high, but not infinite, call it *c.* Let this be agreed upon as a fundamental fact of nature. It then follows that the above-mentioned discrimination between 'before and after' or 'earlier and later' (based on the cause-and-effect relation) is not universally applicable, it breaks down in some cases. This is not as easily explained in non-mathematical language. Not that the mathematical scheme is so complicated. But everyday language is preju- dicial in that it is so thoroughly imbued with the notion of time - you cannot use a verb *(verbum,* 'the' word, Germ. *Zeitwort)* without using it in one or the other tense.

The simplest but, as will turn out, not fully adequate consideration runs thus. Given an event *A.* Contemplate at any later time an event *B* outside the sphere of radius *ct* around *A*. Then *B* cannot exhibit any 'vestige' of *A;* nor, of course can *A* from *B*. Thus our criterion breaks down. By the language we used we have, of course, dubbed *B* to be the later. But are we right in this, since the criterion breaks down either way?

Contemplate at a time earlier (by t) an event *B'* outside that same sphere. In this case, just as before, no vestige of ***B'*** can have reached *A* (and, of course, none from *A* can be exhibited on **B'**).

Thus in both cases there is exactly the same relationship of mutial non-interference. There is no conceptual difference between the classes *B* and *B'* with regard to their cause-effect relation to *A*. So if we want to make this relation, and not a linguistic prejudice, the basis of the 'before and after', then the *B* and *B'* form one class of events that are neither earlier nor later than *A*. The region of space-time occupied by this class is called the region of 'potential simultaneity' (with respect to event A). This expression is used, because a space-time frame can always be adopted that makes *A* simultaneous with a selected particular *B* or a particular *B'*. This was Einstein's discovery (which goes under the name of The Theory of Special Relativity, 1905).

Now these things have become very concrete reality to us

physicists, we use them in everyday work just as we use the multiplication table or Pythagoras' theorem on right-angled triangles. I have sometimes wondered why they made such a great stir both among the general public and among philos- ophers. I suppose it is this, that it meant the dethronement of time as a rigid tyrant imposed on us from outside, a liberation from the unbreakable rule of 'before and after'. For indeed time is our most severe master by ostensibly restricting the existence of each of us to narrow limits - seventy or eighty years, as the Pentateuch has it. To be allowed to play about with such a master's programme believed unassailable until then, to play about with it albeit in a small way, seems to be a great relief, it seems to encourage the thought that the whole 'timetable' is probably not quite as serious as it appears at first sight. And this thought is a religious thought, nay I should call it *the* religious thought.

Einstein has not - as you sometimes hear - given the lie to Kant's deep thoughts on the idealization of space and time; he has, on the contrary, made a large step towards its accom- plishment.

I have spoken of the impact of Plato, Kant and Einstein on the philosophical and religious outlook. Now between Kant and Einstein, about a generation before the latter, physical science had witnessed a momentous event which might have seemed calculated to stir the thoughts of philosophers, men- in-the-street and ladies in the drawing-room at least as much asthe theory of relativity, if not more so. That this was not the case is, I believe, due to the fact that this turn of thought is even more difficult to understand and was therefore grasped by very few among the three categories of persons, at the best by one or another p'.1ilosopher. This event is attached to the names of the American Willard Gibbs and the Austrian Ludwig Boltzmann. I will now say something about it.

With very few exceptions (that really are exceptions) the course of events in nature is irreversible. If we try to imagine a time-sequence of phenomena exactly opposite to one that is actually observed - as in a cinema film projected in reversed order - such a reversed sequence, though it can easily be imagined, would nearly always be in gross contradiction to well-established laws of physical science.

The general 'directedness' of all happening was explained by the mechanical or statistical theory of heat, and this explanation was duly hailed as its most admirable achievement. I cannot enter here on the details of the physical theory, and this is not necessary for grasping the gist of the explana- tion. This would have been very poor, had irreversibility been stuck in as a fundamental property of the microscopic mech- anism of atoms and molecules. This would not have been better than many a medieval purely verbal explanation such as: fire is hot on account of its fiery quality. No. According to Boltzmann we are faced with the natural tendency of any state of order to turn on its own into a less orderly state, but not the other way round. Take as a simile a set of playing cards that you have carefully arranged, beginning with 7, 8, 9, Io, knave, queen, king, ace of hearts, then the same in diamonds, etc. If this well-ordered set is shuffled once, twice or three times it will gradually turn into a random set. But this is not an intrinsic property of the process of shuffling. Given the resulting disorderly set, a process of shuffling is perfectly thinkable that would exactly cancel the effect of the first shuffling and restore the original order. Yet everybody will expect the first course to take place, nobody the second - indeed he might have to wait pretty long for it to happen by chance.

Now this is the gist of Boltzmann's explanation of the unidirectional character of everything that happens in nature (including, of course, the life-history of an organism from birth to death). Its very virtue is that the 'arrow of time' (as Eddington called it) is not worked into the mechanisms of interaction, represented in our simile by the mechanical act of shuffling. This act, this mechanism is as yet innocent of any notion of past and future, it is in itself completely reversible, the 'arrow' - the very notion of past and future - results from statistical considerations. In our simile with the cards the point is this, that there is only one, or a very few, well-ordered arrangements of the cards, but billions of billions of disorderly ones.

Yet the theory has been opposed, again and again, occasionally by very clever people. The opposition boils down to this: the theory is said to be unsound on logical grounds. For, so it is said, if the basic mechanisms do not distinguish between the two directions of time, but work perfectly sym- metrically in this respect, how should there from their co- operation result a behaviour of the whole, an integrated behaviour, that is strongly biased in one direction? Whatever holds for this direction must hold equally well for the opposite one.

If this argument is sound, it seems to be fatal. For it is aimed at the very point which was regarded as the chief virtue of the theory: to derive irreversible events from reversible basic mechanisms.

The argument is perfectly sound, yet it is not fatal. The argument is sound in asserting that what holds for one direction also holds for the opposite direction of time, which from the outset is introduced as a perfectly symmetrical variable. But you must not jump to the conclusion that it holds quite in general for both directions. In the most cautious wording one has to say that in any particular case it holds for either the one or the other direction. To this one must add: in the particular case of the world as we know it, the 'running down' (to use a phrase that has been occasionally adopted) takes place in one direction and this we call the direction from past to future. In other words the statistical theory of heat must be allowed to decide by itself high-handedly, by its own definition, in which direction time flows. (This has a momen- tous consequence for the methodology of the physicist. **He** must never introduce anything that decides independently upon the arrow of time, else Boltzmann's beautiful building collapses.)

It might be feared that in different physical systems the statistical definition of time might not always result in the same time-direction. Boltzmann boldly faced this eventuality; he maintained that if the universe is sufficiently extended and/or exists for a sufficiently long period, time might actually run in the opposite direction in distant parts of the world. The point has been argued, but it is hardly worth while arguing any longer. Boltzmann did not know what to us is at least extremely likely, namely that the universe, as we know it, is neither large enough nor old enough to give rise to such reversions on a large scale. I beg to be allowed to add without detailed explanations that on a very small scale, both in space and in time, such reversions have been observed (Brownian movement, Smoluchowski).

To my view the 'statistical theory of time' has an even stronger bearing on the philosophy of time than the theory of relativity. The latter, however revolutionary, leaves untouched the undirectional flow of time, which it presup- poses, while the statistical theory constructs it from the order of the events. This means a liberation from the tyranny of old Chronos. What we in our minds construct ourselves cannot, so I feel, have dictatorial power over our mind, neither the power of bringing it to the fore nor the power of annihilating it. But some of you, I am sure, will call this mysticism. So with all due acknowledgment to the fact that physical theory is at all times relative, in that it depends on certain basic assump- tions, we may, or so I believe, assert that physical theory in its present stage strongly suggests the indestructibility of Mind by Time.

CHAPTER 6

The Mystery of the Sensual Qualities

In this last chapter I wish to demonstrate in a little more detail the very strange state of affairs already noticed in a famous fragment of Democritus of Abdera - the strange fact that on the one hand all our knowledge about the world around us, both that gained in everyday life and that revealed by the most carefully planned and painstaking laboratory experiments, rests entirely on immediate sense perception, while on the other hand this knowledge fails to reveal the relations of the sense perceptions to the outside world, so that in the picture or model we form of the outside world, guided by our scientific discoveries, all sensual qualities are absent. While the first part of this statement is, so I believe, easily granted by everybody, the second half is perhaps not so frequently realized, simply because the non-scientist has, as a rule, a great reverence for science and credits us scientists with being able, by our 'fabulously refined methods', to make out what, by its very nature, no human can possibly make out and never will be able to make out.

If you ask a physicist what is his idea of yellow light, he will tell you that it is transversal electro-magnetic waves of wave-length in the neighbourhood of 590 millimicrons. If you ask him: But where does yellow come in? he will say: In my picture not at all, but these kinds of vibrations, when they hit the retina of a healthy eye, give the person whose eye it is the sensation of yellow. **On** further inquiry you may hear that different wave-lengths produce different colour-sensations, but not all do so, only those between about 800 and 400 μμ. To the physicist the infra-red (more than 800 μμ) and the ultra-violet (less than 400 μμ) waves are much the same kind of phenomena as those in the region between 800 and 400 μμ, to which the eye is sensitive. How does this peculiar selection come about? It is obviously an adaptation to the sun's radiation, which is strongest in this region of wave-lengths but falls off at either end. Moreover, the intrinsically brightest colour-sensation, the yellow, is encountered at that place (within the said region) where the sun's radiation exhibits its maximum, a true peak.

We may further ask: Is radiation in the neighbourhood of wave-length 590 μμ the only one to produce the sensation of yellow? The answer is: Not at all. If waves of760 μμ, which by themselves produce the sensation of red, are mixed in a definite proportion with waves of 535 μμ, which by themselves produce the sensation of green, this mixture produces a yellow that is indistinguishable from the one produced by 590 μμ. Two adjacent fields illuminated, one by the mixture, the other by the single spectral light, look exactly alike, you cannot tell which is which. Could this be foretold from the wave-lengths - is there a numerical connection with these physical, objective characteristics of the waves? No. Of course, the chart of all mixtures of this kind has been plotted empirically; it is called the colour triangle. But it is not simply connected with the wave-lengths. There is no general rule that a mixture of two spectral lights matches one between them; for example a mixture of 'red' and 'blue' from the extremities of the spectrum gives 'purple', which is not produced by any single spectral light. Moreover, the said chart, the colour triangle, varies slightly from one person to the other, and differs considerably for some persons, called anomalous trichromates (who are *not* colour-blind).

The sensation of colour cannot be accounted for by the physicist's objective picture of light-waves. Could the physiologist account for it, if he had fuller knowledge than he has of the processes in the retina and the nervous processes set up by them in the optical nerve bundles and in the brain? I do not think so. We could at best attain to an objective knowledge of what nerve fibres are excited and in what proportion, perhaps even to know exactly the processes they produce in certain brain cells - whenever your mind registers the sensation of yellow in a particular direction or domain of our field of vision. But even such intimate knowledge would not tell us anything about the sensation of colour, more particularly of yellow in this direction - the same physiological processes might conceivably result in a sensation of sweet taste, or anything else. I mean to say simply this, that we may be sure there is no nervous process whose objective description includes the characteristic 'yellow colour' or 'sweet taste', just as little as the objective description of an electro-magnetic wave includes either of these characteristics.

The same holds for other sensations. It is quite interesting to compare the perception of colour, which we have just surveyed, with that of sound. It is normally conveyed to us by elastic waves of compression and dilatation, propagated in the air. Their wave-length - or *to* be more accurate their frequency - determines the pitch of the sound heard. (**N.B.** The physiological relevance pertains to the frequency, not to the wave-length, also in the case of light, where, however, the two are virtually exact reciprocals of each other, since the velocities of propagation in empty space and in air do not differ perceptibly.) I need not tell you that the range of frequencies of'audible sound' is very different from that of'visible light', it ranges from about 12 or 16 per second to 20,000 or30,000 per second, while those for light are of the order of several hundred (English) billions. The relative range, however, is much wider for sound, it embraces about 10 octaves (against hardly one for 'visible light'); moreover, it changes with the individual, especially with age: the upper limit is regularly and considerably reduced as age advances. But the most striking fact about sound is that a mixture of several distinct frequencies never combines to produce just one intermediate pitch such as could be produced by one intermediate frequency. To a large extent the superposed pitches are perceived separately - though simultaneously - especially by highly musical persons. The admixture of many higher notes ('overtones') of various qualities and intensities results in what is called the timbre (German: *Klangfarbe),* by which we learn to distinguish a violin, a bugle, a church bell, piano ... even from a single note that is sounded. But even noises have their timbre, from which we may infer what is going on; and even my dog is familiar with the peculiar noise of the opening of a certain tin box, out of which he occasionally receives a biscuit. In all this the ratios of the co-operating frequencies are all-important. If they are all changed in the same ratio, as on playing a gramophone record too slow or too fast, you still recognize what is going on. **Yet** some relevant distinctions depend on the absolute frequencies of certain components. If a gramophone record containing a human voice is played too fast, the vowels change perceptibly, in particular the 'a' as in 'car' changes into that in 'care'. A continuous range of frequencies is always disagreeable, whether offered as a sequence, as by a siren or a howling cat, or simultaneously, which is difficult to implement, except perhaps by a host of sirens or a regiment of howling cats. This is again entirely different from the case of light perception. All the colours which we normally perceive are produced by continuous mixtures; and a continuous gradation of tints, in a painting or in nature, is sometimes of great beauty.

The chief characteristics of sound percep!ion are well

understood in the mechanism of the ear, of v/hich we have better and safer knowledge than of the chemistry of the retina. The principal organ is the *cochlea,* a coiled bony tube which resembles the shell of a certain type of sea-snail: a tiny wiuding staircase that gets narrower and narrower as it 'ascends'. In place of the steps (to continue our simile), across the winding staircase elastic fibres are stretched, forming a membrane, the width of the membrane (or the length of the individual fibre) diminishing from the 'bottom' to the 'top'. Thus, like the strings of a harp or a piano, the fibres of different length respond mechanically to oscillations of differ- ent frequency. To a definite frequency a definite small area of the membrane - not just one fibre - responds, to a higher frequency another area, where the fibres are shorter. A mechanical vibration of definite frequency must set up, in each of that group of nerve fibres, the well-known nerve impulses that are propagated to certain regions of the cerebral cortex. \'Ve have the general knowledge that the process of conduction is very much the same in all nerves and changes only with the intensity of excitation; the latter affects the frequency of the pulses, which, of course, must not be confused with the frequency of sound in our case (the two have nothing to do with each other).

The picture is not as simple as we might wish it to be. Had a physicist constructed the ear, with a view to procuring for its owner the incredibly fine discrimination of pitch and timbre that he actually possesses, the physicist would have con- structed it differently. But perhaps he would have come back to it. It would be simpler and nicer if we could say that every single 'string' across the cochlea answers only to one sharply defined frequency of the incoming vibration. This is not so. But why is it not so? Because the vibrations of these 'strings' are strongly damped. This, of necessity, broadens their range of resonance. Our physicist might have constructed them with as little damping as he could manage. But this would have the terrible consequence that the perception of a sound would not cease almost immediately when the producing wave ceases; it would last for some time, until the poorly damped resonator in the cochlea died down. The discrimination of pitch would be obtained by sacrificing the discrimination in time between subsequent sounds. It is puzzling how the actual mechanism manages to reconcile both in a most consummate fashion.

I have gone into some detail here, in order to make you feel that neither the physicist's description, nor that of the physi- ologist, contains any trait of the sensation of sound. Any description of this kind is bound to end with a sentence like: those nerve impulses are conducted to a certain portion of the brain, where they are registered as a sequence of sounds. We can follow the pressure changes in the air as they produce vibrations of the ear-drum, we can see how its motion is transferred by a chain of tiny bones to another membrane and eventually to parts of the membrane inside the cochlea, composed of fibres of varying length, described above. We may reach an understanding of how such a vibrating fibre sets up an electrical and chemical process of conduction in the nervous fibre with which it is in touch. We may follow this conduction to the cerebral cortex and we may even obtain some objective knowledge of some of the things that happen there. But nowhere shall we hit on this 'registering as sound', which simply is not contained in our scientific picture, but is only in the mind of the person whose ear and brain we are speaking of.

We could discuss in similar manner the sensations of touch, of hot and cold, of smell and of taste. The latter two, the chemical senses as they are sometimes called (smell affording an examination of gaseous stuffs, taste that of fluids), have this in common with the visual sensation, that to an infinite number of possible stimuli they respond with a restricted manifold of sensate qualities, in the case of taste: bitter, sweet, sour and salty and their peculiar mixtures. Smell is, I believe, more various than taste, and particularly in certain animals it is much more refined than in man. What objective features of a physical or chemical stimulus modify the sensation notice- ably seems to vary greatly in the animal kingdom. Bees, for instance, have a colour vision reaching well into the ultra- violet; they are true trichromates (not dichromates, as they seemed in earlier experiments which paid no attention to the ultra-violet). It is of very particular interest that bees, as von Frisch in Munich found out not long ago, are peculiarly sensitive to traces of polarization of light; this aids their orientation with respect to the sun in a puzzlingly elaborate way. To a human being even completely polarized light is indistinguishable from ordinary, non-polarized light. Bats have been discovered to be sensible to extremely high fre- quency vibrations ('ultra-sound') far beyond the upper limit of human audition; they produce it themselves, using it as a sort of 'radar', to avoid obstacles. The human sense of hot or cold exhibits the queer feature of ''les extremes se touchent': if we inadvertently touch a very cold object, we may for a moment believe that it is hot and has burnt our fingers.

Some twenty or thirty years ago chemists in the U.S.A. discovered a curious compound, of which I have forgotten the chemical name, a white powder, that is tasteless to some persons, but intensely bitter to others. This fact has aroused keen interest and has been widely investigated since. The quality of being a 'taster' (for this particular substance) is inherent in the individual, irrespective of any other condi- tions. Moreover, it is inherited according to the Mendel laws in a way familiar from the inheritance of blood group characteristics.Just as with the latter, there appears to be no conceivable advantage or disadvantage implied by your being a 'taster' or a 'non-taster'. One of the two 'alleles' is dominant in heterozygotes, I believe it is that of the taster. It seems to me very improbable that this substance, discovered haphaz- ardly, should be unique. Very probably 'tastes differ' in quite a general way, and in a very real sense!

Let us now return to the case of light and probe a little deeper into the way it is produced and into the fashion in which the physicist makes out its objective characteristics. I suppose that by now it is common knowledge that light is usually produced by electrons, in particular by those in an atom where they 'do something' around the nucleus. **An** electron is neither red nor blue nor any other colour; the same holds for the proton, the nucleus of the hydrogen atom. But the union of the two in the atom of hydrogen, according to the physicist, produces electro-magnetic radiation of a certain discrete array of wave-lengths. The homogeneous constituents of this radiation, when separated by a prism or an optical grating, stimulate in an observer the sensations of red, green, blue, violet by the intermediary of certain physiological processes, whose general character is sufficiently well known to assert that they are not red or green or blue, in fact that the nervous elements in question display no colour in virtue of their being stimulated; the white or grey the nerve cells exhibit whether stimulated or not is certainly insignificant in respect of the colour sensation which, in the individual whose nerves they are, accompanies their excitation.

Yet our knowledge of the radiation of the hydrogen atom and of the objective, physical properties of this radiation originated from someone's observing those coloured spectral lines in certain positions within the spectrum obtained from glowing hydrogen vapour. This procured the first knowledge, but by no means the complete knowledge. To achieve it, the elimination of the sensates has to set in at once, and is worth pursuing in this characteristic example. The colour in itself tells you nothing about the wave-length; in fact we have seen before that, for example, a yellow spectral line might conceiv- ably be not 'monochromatic' in the physicist's sense, but composed of many different wave-lengths, if we did not know that the construction of our spectroscope excludes this. It gathers light of a definite wave-length at a definite position in the spectrum. The light appearing there has always exactly the same colour from whatever source it stems. Even so the quality of the colour sensation gives no direct clue whatsoever to infer the physical property, the wave-length, and that quite apart from the comparative poorness of our discrimination of hues, which would not satisfy the physicist. *A priori* the sensation of blue might conceivably be stimulated by long waves and that of red by short waves, instead of the other way round, as it is.

To complete our knowledge of the physical properties of the light coming from any source a special kind of spectroscope has to be used; the decomposition is achieved by a diffraction grating. A prism would not do, because you do not know beforehand the angles under which it refracts the different wave-lengths. They are different for prisms of different material. In fact, *a priori*, with a prism you could not even tell that the more strongly deviated radiation is of shorter wave-length, as is actually the case.

The theory of the diffraction grating is much simpler than that of a prism. From the basic physical assumption about light - merely that it is a wave phenomenon - you can, if you have measured the number of the equidistant furrows of the grating per inch (usually of the order of many thousands), tell the exact angle of deviation for a given wave-length, and therefore, inversely, you can infer the wave-length from the 'grating constant' and the angle of deviation. In some cases (notably in the Zeeman and Stark effects) some of the spectral lines are polarized. To complete the physical description in this respect, in which the human eye is entirely insensitive, you put a polarizer (a Nicol prism) in the path of the beam, before decomposing it; on slowly rotating the Nicol around its axis certaii;i lines are extinguished or reduced to minimal brightness for certain orientations of the Nicol, which indicate the direction (orthogonal to the beam) of their total or partial polarization.

Once this whole technique is developed, it can be extended far beyond the visible region. The spectral lines of glowing vapours are by no means restricted to the visible region, which is not distinguished physically. The lines form long, theoret- ically infinite series. The wave-lengths of each series are connected by a relatively simple mathematical law, peculiar to it, that holds uniformly throughout the series with no distinction of that part of the series that happens to lie in the visible region. These serial laws were first found empirically, but are now understood theoretically. Naturally, outside the visible region a photographic plate has to replace the eye. The wave-lengths are inferred from pure measurements of lengths: first, once and for all, of the grating constant, that is the distance between neighbouring furrows (the reciprocal of the number of furrows per unit length), then by measuring the positions of the lines on the photographic plate, from which, together with the known dimensions of the apparatus, the angles of deviation can be computed.

These are well-known things, but I wish to stress two

points of general importance, which apply to well-nigh every phys- ical measurement.

The state of affairs on which I have enlarged here at some length is often described by saying that, as the technique of measuring is refined, the observer is gradually replaced by more and more elaborate apparatus. Now this is, certainly in the present case, not true; he is not gradually replaced, but is so from the outset. I tried to explain that the observer's colourful impression of the phenomenon vouchsafes not the slightest clue to its physical nature. The device of ruling a grating and measuring certain lengths and angles has to be introduced, before even the roughest qualitative knowledge of what we call the objective physical nature of the light and of its physical components can be obtained. And this is the relevant step. That the device is later on gradually refined, while remaining essentially always the same, is epistemologically unimportant, however great the improve- ment achieved.

The second point is that the observer is never entirely replaced by instruments; for if he were, he could obviously obtain no knowledge whatsoever. **He** must have constructed the instrument and, either while constructing it or after, he must have made careful measurements of its dimensions and checks on its moving parts (say a supporting arm turning around a conical pin and sliding along a circular scale of angles) in order to ascertain that the movement is exactly the intended one. True, for some of these measurements and check-ups the physicist will depend on the factory that has produced and delivered the instrument; still all this informa- tion goes back ultimately to the sense perceptions of some living person or persons, however many ingenious devices may have been used to facilitate the labour. *Finally* the observer must, in using the instrument for his investigation, take readings on it, be they direct readings of angles or of distances, measured under the microscope, or between spec- tral lines recorded on a photographic plate. Many helpful devices can facilitate this work, for instance photometric recording across the plate of its transparency, which yields a magnified diagram on which the positions of the lines can be easily read. But they must be read! The observer's senses have to step in eventually. The most careful record, when not inspected, tells us nothing.

So we come back to this strange state of affairs. While 1he direct sensual perception of the phenomenon tells us nothing as to its objective physical nature (or what we usually call so) and has to be discarded from the outset as a source of information, yet the theoretical picture we obtain eventually rests entirely on a complicated array of various informations, all obtained by direct sensual perception. It resides upon them, it is pieced together from them, yet it cannot really be said to contain them. In using the picture we usually forget about them, except in the quite general way that we know our idea of a light-wave is not a haphazard invention of a crank but is based on experiment.

I was surprised when I discovered for myself that this state of affairs was clearly understood by the great Democritus in the fifth century B.c., who had no knowledge of any physical measuring devices remotely comparable to those I have been telling you about (which are of the simplest used in our time).

Galen us has preserved us a fragment (Diels, fr. r 25), in which Democritus introduces the intellect (ouivota) having an argument with the senses (aicr0tjcn:v;) about what is 'real'. The former says: 'Ostensibly there is colour, ostensibly sweet- ness, ostensibly bitterness, actually only atoms and the void', to which the senses retort: 'Poor intellect, do you hope to defeat us while from us you borrow your evidence? Your victory is your defeat.'

In this chapter I have tried by simple examples, taken from the humblest of sciences, namely physics, to contrast the two general facts (a) that all scientific knowledge is based on sense perception, and (b) that none the less the scientific views of natural processes formed in this way lack all sensual qualities and therefore cannot account for the latter. Let me conclude with a general remark.

Scientific theories serve to facilitate the survey of our observations and experimental findings. Every scientist knows how difficult it is to remember a moderately extended group of facts, before at least some primitive theoretical picture about them has been shaped. It is therefore small wonder, and by no means to be blamed on the authors of original papers or of text-books, that after a reasonably coherent theory has been formed, they do not describe the bare facts they have found or wish to convey to the reader, but clothe them in the termin- ology of that theory or theories. This procedure, while very useful for our remembering the facts in a well-ordered pattern, tends to obliterate the distinction between the actual observations and the theory arisen from them. And since the former always are of some sensual quality, theories are easily thought to account for sensual qualities; which, of course, they never do.

AUTOBIOGRAPHICAL SKETCHES

I lived far apart from my best friend, actually the only close friend I ever had, for the greater part of my life. (Maybe that is why I have often been accused of flirtatiousness instead of true friendship.) He studied biology (botany to be exact); I physics. And many a night we would stroll back and forth between Gluckgasse and Schli.isselgasse engrossed in philo- sophical conversation. Little did we know then that what seemed original to us had occupied great minds for centuries already. Don't teachers always do their best to avoid these topics for fear that they might conflict with religious doctrines and cause uncomfortable questions? This is the main reason for my turning against religion, which has never done me any harm.

I am not sure whether it was right after the First World

War or during the time I spent in Zurich (1921-7) or even later in Berlin (1927-33) that Franzel and I spent a long evening together again. The small hours of the morning found us still talking in a cafe on the outskirts of Vienna. He seemed to have changed a lot with the years. After all, our letters had been few and far between and of very little substance.

I might have added earlier that we also spent our time together reading Richard Semon. Never before or after did I read a serious book with anyone else. Richard Semon was soon banned by the biologists, since his views, as they saw them, were based on the inheritance of acquired characteris- tics. So his name was forgotten. Many years later I encoun- tered him in a book *(Human Knowledge?)* by Bertrand Russell, who devoted a thorough study to this genial biologist, stress- ing the significance of his Mneme theory.

Franzel and I did not see each other again until 1956. This time it was a very brief encounter in our flat in Vienna, Pasteurgasse 4, while others were present, so that those fifteen minutes are hardly worth mentioning. Franzel and his wife lived across the border, our northern one, unhampered by the authorities, it seemed; nevertheless, leaving the country had become rather difficult. We never met again: two years later he died very suddenly.

Today I am still friends with his charming nephew and niece, his favourite brother Silvio's children. Silvio, the youngest in the family, was a doctor in Krems, where I went to see him when I returned to Austria in 1956. He must have been seriously ill already, for he died not long afterwards. One of Franzel's brothers, **E.,** is still alive. He is a respected surgeon **in** Klagenfurt. **E.** once took me up the Einser (Sextener Dolomites) and, what's more, saw me safely down again. I am afraid we have lost contact, driven apart by our different views of the world.

Shortly before I entered the University ofVienna **in** 1906, the only university I was ever enrolled in, the great Ludwig Boltzmann met his sad end in Duino. To this day I have not forgotten the clear, precise and yet still enthusiastic words with which Fritz Hasenohrl described Boltzmann's work to us. Boltzmann's scholar and successor held his inaugural address in autumn 1907 in the primitive lecture hall of the old Türken- strasse building without any pomp or ceremony. I was deeply impressed by his introduction, and no perception in physics has ever seemed more important to me than that of Boltzmann - despite Planck and Einstein. Incidentally, Einstein's early work (before 1905) shows how fascinated he too was by Boltz- mann's work. **He** was the only one who took a major step

beyond it by inverting Boltzmann's equation $S = k \lg W$. No

other human being had a greater influence on me than Fritz Hasenohrl - except perhaps my father Rudolph, who in the course of those many years we lived together drew me into conversations concerning his many interests. But more about that later.

While still a student I made friends with Hans Thirring.

This turned out to be a lasting relationship. When Hasenohrl was killed in action in 19 r 6, Hans Thirring became his successor; he retired at seventy, forgoing the privilege of remaining for the honorary year and leaving Boltzmann's professorial chair to his son, Walter.

After 1911, while I was assistant to Exner, I met K. W. F. Kohlrausch, and yet another lasting friendship began. Kohl- rausch had made his name by proving experimentally the existence of the so-called 'Schweidle Fluctuations'. In the year before the outbreak of the war we worked together on the research of 'secondary radiations', which produced - at the smallest possible angle on small plates of varying material - a (mixed) beam of gamma rays. I learnt two things in those years: firstly that I was not suited to experimental work, and secondly that my surroundings and the people who were part of them were no longer capable of making experimental progress on a big scale. There were many reasons for this, one of them being that in charming old Vienna well-meaning blunderers were placed, often according to seniority, in key positions, thus impeding all progress. If only it had been realized that per- sonalities with great mental capacities were needed, even if it meant bringing them in from afar! The theories of atmospheric electricity and radio activity were both originally developed in Vienna, but anyone who felt really dedicated to their work had to follow those theories wherever they had been passed on. Lise Meitner, for instance, left Vienna and went to Berlin.

But back to myself: **in** retrospect I am very grateful that because of my reserve officer's training in 1910/I I I was appointed assistant to Fritz Exner and not to Hasenohrl. It meant that I was able to experiment with **K. W. F.** Kohlrausch and make use of a number of beautiful instruments, take them to my room, especially the optical ones, and dabble with them to my heart's content. Thus I could set the interferometer, admire the spectra, mix colours, etc. This was also how I discovered - through the Rayleigh equation - the deuter anomaly ofmy eyes. Moreover I was committed to do the long practical course, so that I learnt to appreciate the significance of measuring. I wish there were more theoretical physicists who did.

In 1918 we had a kind of revolution. The Emperor Karl abdicated and Austria became a republic. Our everyday life remained much the same. However, my life was affected by the breaking up of the Empire. I had accepted a post as a lecturer in theoretical physics in Czernowitz and had already envisaged spending all my free time acquiring a deeper knowledge of philosophy, having just discovered Schopenhauer, who intro- duced me to the Unified Theory of the Upanishads.

Forus Viennese the war and its consequences meant that we could no longer satisfy our basic needs. Hunger was the punishment the victorious Entente had chosen in retaliation for the unlimited U-boat war of their enemies, a war so atrocious that Prince Bismarck's heir and his followers could only outdo it in quantity, and not in quality, in the Second World War. Hunger prevailed throughout the country except on the farms, where our poor women were sent to ask for eggs, butter and milk. Despite the goods with which they paid - knitted garments, pretty petticoats, etc. - they were sneered at and treated like beggars.

In Vienna it had become virtually impossible to socialize and entertain friends. There was simply nothing to offer, and even the simplest dishes were reserved for Sunday lunch. In some ways this lack of social activities was compensated by the daily visit to the community kitchens. The *Gemein- schaftskiichen* were often referred to as *Gemeinheitskiichen* (*Gemein-*

schaft = 'community'; *Gemeinheit* = 'a mean trick'). There

we met for lunch. We had to be grateful to the women who considered it their responsibility to create meals out of nothing. It is no doubt easier to do this for 30 or 50 people than for three. Besides, relieving others of a burden must in itself be rewarding.

My parents and I met a number of people with similar interests there and some of them, the Radons, for example, both of them mathematicians, became great friends of our family.

I believe that in one way my parents and I were particularly disadvantaged. At that time we lived in a large flat (actually two flats made into one) on the fifth floor of a rather valuable building in the city, which belonged to my mother's father. It had no electric light, partly because my grandfather did not want to pay for having it installed and also because my father, in particular, had become so used to the excellent gas light at a time when light bulbs were still very expensive and ineffi- cient that we really saw no need for them. And we had the old tiled stoves removed and replaced by solid gas stoves with copper reflectors - servants were hard to come by in those days, and we had hoped to make things easier for ourselves. Gas was also used for cooking, although we did still have an enormous old wood-burning stove standing in the kitchen. This was all very well until one day one of the higher bureaucratic offices, probably the city council, decreed that gas was to be rationed. From that day on every household was allowed one cubic metre per day regardless of how the fuel had to be used. If anyone was found using more, they were simply cut off.

In the summer of 1919 we went to Millstadt, Carinthia, and my father, who was sixty-two, showed the first signs of ageing and of what was to be his final illness, a fact we did not become aware of at the time. Whenever we went for a walk he would lag behind, especially where it got steep, and he would feign botanical curiosity to mask his exhaustion. From about 1902 on Father's main interest was botany. During the summer months he collected material for his studies, not for setting up a herbarium of his own, but for experimenting with his microscope and microtome. He had become a mor- phogeneticist and phylogeneticist and had abandoned his dedication to Italy's great painters and also his own artistic interests, which consisted of sketching innumerable land- scapes. Father;s rather bored reaction to our coaxing: **'Oh,** Rudolph, do come on' and **'Mr** Schrodinger, it's getting rather late', did not alarm us either; we were actually used to that; so we put it down to his absorbed concentration.

After our return to Vienna the signs became more apparent, but still we did not take them seriously as a warning: frequent and heavy bleeding from his nose and retina, and finally fluid in his legs. I think he knew long before everyone else that his end was near. Unfortunately this was just the time of the gas calamity mentioned above. We acquired carbon lamps, and he insisted on tending them himself. A dreadful stench spread from his beautiful library, which he had turned into a carbide laboratory. Twenty years earlier, when he had learnt to etch with Schmutzer, he had used the room to soak his copper and zinc plates in acids and chlorinated water; I was still at school then, and had shown great interest in his activities. But now I left him to his own devices. I was glad to be back at my beloved physics institute after serving in the war for almost four years. Besides, in autumn 1919 I became engaged to the girl who has been my wife for forty years now. I do not know whether my father had adequate medical treatment, but what I do kriow is that I should have looked after him better. I should have asked Richard von Wettstein, who was after all a good friend of his, to seek help at the medical faculty. Would better advice have slowed down his arteriosclerosis? And if so, would it have been to the advantage of a sick man? Only Father was fully aware of our financial situation after the closing down of our oilcloth and linoleum store on the Stephansplatz in 1917 (due to lack of stock).

He died peacefully on Christmas Eve 1919, in his old armchair.

The following year was that of rampant inflation, which meant the depreciation of Father's meagre bank account, which would never have kept my parents' heads above water anyway. The proceeds of the Persian rugs he had sold (with my consent!) dissolved into nothing; gone for ever were the microscopes, the microtome and a good part of his library, which I gave away for a song after his death. His greatest worry during the last months had been that at the ripe old age of thirty-two I was earning virtually nothing - 1 ,ooo Austrian kronen (before tax, that is, for I am sure he listed it in his tax declaration except when I was an officer during the war). The only success of his son that he lived to see was that I had been offered (and had also accepted) a better-paid post as private lecturer and assistant to Max Wien in Jena.

My wife and I moved to Jena in April 1920, leaving my mother to fend for herself, in fact which I am not at all proud of today. She had to bear the burden of packing and clearing the flat. Oh, how blind we all were! Her father, who owned the house, was rather worried after my father's death about who would pay the rent. We were in no position to do so, and Mother had to make room for a more affluent tenant. My future father-in-law kindly turned up with the man, a Jewish businessman working for the Phoenix, a prosperous insurance company. So Mother had to leave, where to I do not know. Had we not been so blind we would have foreseen - and thousands of similar cases would have proved us right - what an excellent source of money the big, well-furnished flat could have proved for my mother had she lived longer. She died in the autumn of

192 1 of cancer of the spine after what we believed had been a successful operation on her breast cancer in 191 7.

I rarely remember dreams, and I seldom had nasty ones -

except maybe in my early childhood. For a long time after my father's death, however, a nightmare kept recurring again and again: my father was still alive and I knew I had given away all his beautiful instruments and botanical books. What was he to do now that I had rashly and irretrievably destroyed the basis of his intellectual life? I am sure it was my guilty conscience that caused the dream, as I had cared so little for my parents between 1919 and1921. This can be the only explanation, as I am not normally bothered with nightmares or a guilty con- science either.

My childhood and adolescence (1887-19 roor thereabouts) was mainly influenced by my father, not in the usual educa- tional manner, but in a more ordinary way. This was due to his spending a lot more time at home than most men who work for a living and to my being at home, too. In my early years of learning I was taught by a private teacher who came to see me twice a week, and at grammar school we still had the blessed tradition of attending for twenty-five hours a week, mornings only. (On two afternoons only we had to attend for protestant religious education.)

I learnt a great deal on those occasions, although the result was not always related to the subject of religion. Time limita- tions concerning school commitments are a great asset. If a pupil feels inclined, he has time for thinking, and he can also take private lessons in the subjects which are not part of the curriculum. I can only find words of praise for my old school (Akademisches Gymnasium): I was rarely bored there, and when it did happen (our preparatory philosophical course was really bad), I would turn my attention to some other subject, my French translation, for example.

At this point I should like to add a remark of a more general kind. The discovery of chromosomes as the decisive factors in heredity seems to have given society the right to overlook other better-known but equally important factors such as communication, education and tradition. It is assumed that these were not so important because from the point of view of genetics they are not stable enough. This is quite true. However, there are cases such as that of Kaspar Hauser, for example, and that of a small group of Tasmanian 'Stone Age' children who were only recently brought to live in English surroundings and granted a first-class English upbringing, with the effect that they reached the educational level of upper-class Englishmen. Does this not prove to us that it takes both a code of chromosomes and civilized human surround- ings to produce people of our kind? In other words, the intellectual level of every individual is bred by 'nature' and by 'nurture'. Schools are therefore (not as our Empress Maria Theresa liked to see it) invaluable for human guidance, and much less for political purposes. And a sound family back- ground is just as important for preparing the soil for the seed the schools will sow. This is unfortunately a fact overlooked by those who claim that only the children of the less educated should attend schools for higher education (will their children be excluded for the same reasons?) and also by British High Society, where it is deemed upper class to replace family life by boarding school and considered a sign of nobility to leave home early. So even the present Queen had to part with her first-born and send him to such an institution. None of this is strictly speaking any of my concern. It only came to my mind when I once again realized how much I gained from the time I spent with my father as a young boy and how little I would have profited from school had he not been there. He actually knew far more than they had to offer, not because he had been forced to study it thirty years earlier, but because he was still interested. If I went into detail here, I should end up telling a long story.

Later on, when he had taken up botany and I had virtually devoured *The Origin of Species,* our discussions took on a different character, certainly different from that conveyed at school, where the theory of evolution was still banned from our biology lessons and teachers of religious education were advised to call it heresy. Of course I soon became an ardent follower of Darwinism (and still am today, for that matter), while Father, influenced by his friends, urged caution. The link between natural selection and the survival of the fittest on the one hand and Mendel's law and de Vries's theory of mutation on the other had yet to be fully discovered. Even today I don't know why zoologists have always tended to swear by Darwin, while botanists appear to be rather more reticent. However, one thing we all agreed on - and when I say 'all',

I particularly remember Hofrat Anton Handlisch, who was a zoologist at the museum of natural history and the one I knew and liked best of all my father's friends - we were all unanimous in holding that the basis of evolutionary theory was causal rather than finalistic; and that no special laws of nature, such as vis viva, or an entelechy, or a force of orthogenesis, etc., were at work in living organisms to abro- gate or to counteract the universal laws of inanimate matter. My religious teacher would not have been happy about this view, but he did not concern me anyway.

Our family was accustomed to travelling in the summer.

This not only brightened my life, but also helped whet my intellectual appetite. I remember one visit to England a year before I started intermediate school *(Mittelschule),* when I stayed with relatives of my mother at Ramsgate. The long, wide beach was ideally suited for donkey rides and learning to handle a bicycle. The strong tidal changes claimed my full attention. Little bathing huts on wheels were set up along the beach, and a man and his horse were always busy moving these cabins up or down according to the tide. On the Channel I first noticed that one could make out the funnel smoke of distant boats on the horizon long before they themselves appeared, a result of the curvature of the water- surface.

In Leamington I met my great-grandmother at Madeira Villa, and as she was called Russell and the street she lived in was called 'Russell', I was convinced it was named after my late great-grandfather. An aunt of my mother's also lived there with her husband, Alfred Kirk, and six Angora cats. (In later years there were said to be twenty.) In addition she had an ordinary tomcat who would very often come home from his nocturnal adventures in a sad state, so he was given the name Thomas Becket (referring to the Archbishop of Canterbury who was killed in office by order of King Henry II) - not that this meant a great deal to me then, nor was it very appro- priate.

It is thanks to my Aunt Minnie, Mother's youngest sister,

who moved from Leamington to Vienna when I was five, that I learnt to speak fluent English long before I could write in German, let alone English. When I was finally introduced to the spelling and reading of the language I thought I knew so well, I was in for a surprise. It was thanks to my mother that half-days of English practice were launched. I was *nat* too pleased about that at the time. We would walk from the Weiherburg down to the pretty and in those years still quiet little town of Innsbruck together, and Mother would say: 'Now we are going to speak English to each other the whole way- not another word of German.' And that is just what we did. I only realized later how much I profited from it to this day. Though forced to leave the country of my birth, I never felt a stranger abroad.

I seem to remember visiting Kenilworth and Warwick on our bicycle tours round Leamington. And on the way back to Innsbruck from England I remember seeing Bruges, Cologne, Coblenz - a steamboat took us up the Rhine - I remember Riidesheim, Frankfurt, Munich, I think; then Innsbruck. I can recall the little boarding house which belonged to Richard Attlmayr.

From there I went to school for the first time, down to St Nikolaus, where I had private tuition, as my parents were afraid I had forgotten my ABC and my sums during the holiday and would fail my entrance exam in the autumn. In later years we nearly always went to the South Tyrol or Carinthia, and sometimes we would go to Venice for a few days in September. There is no end to the list of beautiful things I was given the chance to see in those days, things that no longer exist, due to the motor car, 'development' and new borders. I think few people then, let alone today, experienced such a happy childhood and adolescence as I did, even though I was an only child. Everyone was friendly towards me and we were all on good terms with each other. If only all teachers, including parents, would take to heart the necessity of mutual understanding! We cannot exert any lasting influence over those entrusted to us without it.

Maybe I ought to say something about my years at university between r 906 and r gr o, as there might not be any chance of doing so later on. I mentioned earlier that Hasenohrl and his carefully conceived four-year course (five hours a week!) influenced me more than anything else. Unfortunately I missed the last year (r g IO/ r r), as I could no longer postpone my national service. As it turned out this was not quite as unpleasant as I had anticipated, for I was sent to the beautiful old town of Cracow and I also spent a memor- able summer near the Carinthian border (near Malborghet). Apart from Hasenohrl's, I attended all the other mathemat- ics lectures I could. Gustav Kohn gave his talks on projective geometry. His style, so severe and clear, left a lasting impres- sion. Kohn would alternate from a pure synthetic method one year - without any formulas - to an analytical approach the next. There is in fact no better example for the existence of axiomatic systems. Through him duality in particular turned out to be a breathtaking phenomenon, differing somewhat in two- and three-dimensional geometry. He also proved to us the profound influence of Felix Klein's group theory on the development of mathematics. The fact that the existence of a fourth harmonic element has to be accepted as an axiom in a two-dimensional structure while it can easily be proved in a three-dimensional was to him the simplest illustration of Goedel's great theorem. There were so many things I learnt from Kohn which I would never have had the time to learn later on.

In my first year I also did qualitative chemical analysis, and certainly gained a lot from it. Skraup's lectures on inorganic chemical analysis were rather good; those on organic chemical analysis, which I read during the summer term, poor in comparison. They could have been ten times as good and still they would hardly have improved my understanding of nucleic acids, enzymes, antibodies and the like. As it was I could only feel my way ahead, led by intuition, which was none the less productive.

On 31 July 1914 my father turned up at my little office in

the Boltzmanngasse to break the news that I had been called up. The Predilsattel in Carinthia was to be my first destination. We went off to buy two guns, a small one and a large one. Fortunately I was never forced to use them on either man or animal, and in 1938 during a search of my flat in Graz I handed them over to the good-natured official,just to be sure.

A few words about the war itself: my first posting, Predilsat- tel, was uneventful. Once, though, we had a false alarm. Our commanding officer, Captain Reindl, had arranged with confi- dants that in the event ofl talian troops advancing up the wide valley towards the lake (Raiblersee), we were to be warned by smoke signals. It so happened that someone was baking pota- toes or burning weeds just along the border. We were told to man the two watch posts and I was put in charge of the one on the left. We spent ten days up there before someone remem- bered to call us back down. Up there I learnt that springy floorboards (with only a sleeping-bag and blanket) are much more comfortable to sleep on than a solid floor. My other observation was of a different nature, something I never came across before or after. One night the guard on duty woke me up to report that he could see a number of lights moving up the slope opposite us, obviously heading toward our position. (Incidentally, this part of the mountain (Seekopf) had no paths at all.) I got out ofmy sleeping-bag and made my way through the connecting passage to the post to take a closer look. The guard was right about the lights, but they were St Elmo's fire on the top of our own wire abatis a couple of yards away, and the displacement against the background was only parallactic. This was because the observer himself was moving. When I stepped out of our spacious dug-out at night I would watch these pretty little fires on the tips of the grass that covered the roof. This was the only time I came across the phenomenon.

After spending much idle time there I was posted to Fran- zensfeste, then to Krems and then to Komorn. For a short time I had to serve at the front. I joined a small unit first at Gorizia, then at Duino. They were equipped with an odd naval gun. We eventually retired to Sistiana, and from there I was sent to a rather boring but none the less beautiful observation post near Prosecco, goo feet above Trieste, where we had an even odder gun. My future wife Annemarie came to see me there, and on one occasion Prince Sixtus of Bourbon, the brother of the Empress Zita, visited our positions. He was not in uniform, and later I learnt that he was in fact our enemy as he was serving in the Belgian army. The reason for this was that the French did not allow any member of the Bourbon family to join their army. The aim of his visit at the time was to bring about a separate peace agreement between Austria-Hungary and the Entente Cordiale, which, of course, meant high treason against Ger- many. Unfortunately his plan never materialized.

My first encounter with Einstein's theory of 1916 was at

Prosecco. I had so much time at my disposal, yet had great difficulties in understanding it. Nevertheless a number of marginal notes I made then still appear reasonably intelligent to me even now. As a rule Einstein would present a new theory in an unnecessarily complicated form, and never more so than in 1945, when he introduced the so-called 'asymmetric' unitary field theory. But perhaps that is not just characteristic of that great man, but nearly always happens when someone postu- lates a new idea. In the case of the above-mentioned theory Pauli told him there and then that it was unnecessary to introduce the complex quantities, because each of his tensor equations consisted of both a symmetric and a sheer symme- tric part anyway. Only in 1952, in an article he wrote together with Mme B. Kaufman for a volume published to celebrate Louis de Broglie's sixtieth birthday, did he agree with my much simpler version by ingeniously excluding the so-called 'strong' version. This was a very important move indeed.

The last year or so of the war I spent as a 'meteorologist' first in Vienna, then Villach, then Wiener Neustadt and finally in Vienna again. This was a great asset to me, as I was spared the disastrous retreat of our badly torn front lines.

In March/ April 1920 Annemarie and I got married. We moved soon after to Jena, where we took furnished lodgings. I was expected to add some up-to-date theoretical physics to Professor Auerbach's set lectures. We enjoyed the friendship and cordiality of both the Auerbachs, who were Jews, and of my boss Max Wien and his wife (they were anti-Semites by tradition, but bore no personal malice). Being on such good terms with them all was a great help to me. In 1933, the Auerbachs, I am told, saw no means of escape from the oppression and humiliation which Hitler's taking over *(Machtergreifung)* held in store for them but suicide. Eberhard Buchwald, a young physicist who had just lost his wife, and a couple called Eller with their two little sons were also amongst our friends in Jena. Mrs Eller came to see me here in Alpbach last summer (1959), a poor bereaved woman whose three men-folk had lost their lives fighting for a cause they did not believe in.

A chronological account of someone's life is one of the most boring things I can think of. Whether you are recalling incidents of your own life or that of someone else, you will rarely find more than the occasional experience or observation worth recounting - even if the historical order of events seems important to you at the time. That is why I am now going to give a short summary of the periods of my life, so that I can refer to them later without having to watch the chronological order.

The first period (1887-1920) ends with my marrying Annemarie and leaving Germany. I shall call it my first Viennese Period. The second period (1920-7) I shall call 'My First Years of Roaming', as I was taken to Jena, Stuttgart, Breslau and finally to Zurich (in 1921). This period ends with my call to Berlin as Max Planck's successor. I had discovered wave mechanics during my stay in Arosa in 1925. My paper had been published in 1926. As a result of this I went on a two-month lecturing tour of North America, which prohibition had dried up successfully. The third period

(1927-33) was a rather nice one. I shall call it 'My Teaching and Learning'. It ended with Hitler's assumption of power, the so-called *Machter- greifung,* in 1933. While completing the summer term of that year I was already busy sending my belongings to Switzerland. At the end ofj uly I left Berlin to spend my holidays in the South Tyrol. The South Tyrol had become Italian under the Treaty of St Germain, so it was still accessible to us with our German passports, whereas Austria was not. Prinz Bismarck's great successor had succeeded in imposing a blockade in Austria which became known as the *Tausendmarksperre.* (My wife, for instance, could not visit her mother on her seventieth birthday. His Excellency's authorities did not give her permission). I did not go back to Berlin after the summer, but instead handed in my resignation, which remained unanswered for a long time. In fact they then denied ever having received it, and when they learnt I had been awarded the Nobel Prize for physics, they flatly refused to accept it.

The fourth period (1933-9) I shall call 'My Later Years of

Roaming'. As early as spring 1933 F. A. Lindemann (later Lord Cherwell) offered me a 'living' in Oxford. This was on the occasion of his first visit to Berlin, when I happened to mention my distaste for the present situation. He faithfully kept his word. And so my wife and I took to the road in a little BMW acquired for the occasion. We left Malcesine and via Bergamo, Lecco, St Gotthard, Zurich and then Paris we reached Brussels, where a Solvay Congress was being held. From there we went to Oxford; we did not travel together. Lindemann had already taken the necessary steps to make me a fellow of Magdalen College, though I received the greater part of my pay from ICI.

When, in 1936, I was offered a chair at Edinburgh Univer- sity and another at Graz, I chose the latter, an extremely foolish thing to do. Both the choice and the outcome were unexampled, though the outcome was a lucky one. Of course I was more or less undermined by the Nazis in 1938, but by then I had already accepted a call to Dublin, where de Valera was about to found the Institute for Advanced Studies. Loyalty towards his own university would never have allowed Edinburgh's **E.T.** Whittaker, de Valera's former teacher, to suggest me for the post had I gone to Edinburgh in 1936. As it was, Max Born was appointed in my stead. Dublin proved a hundred times better for me. Not only would the work in Edinburgh have been a great burden to me, but so would the position of enemy alien in Great Britain throughout the war.

Our second 'escape' took us from Graz, via Rome, Geneva and Zurich to Oxford where our dear friends, the Whiteheads, put us up for two months. This time we had to leave our good little **BMW,** 'Grauling', behind, as it would have been too slow, and besides, I no longer possessed a driving licence. The Dublin Institute was not yet 'ready', and so my wife, Hilde, Ruth and I went to Belgium in December 1938. First I held lectures (in German!) at the University of Ghent as guest professor; this was for the 'fondation Franqui-Seminar'. Later on we spent about four months in Lapanne by the sea. It was a lovely time - despite the jellyfish. It was also the only time I ever came across the phosphorescence of the sea. In Septem- ber 1939, the first month of the Second World War, we left for Dublin via England. With our German passports we were still enemy aliens to the British, but obviously thanks to de Valera's letters of reference we were granted transit. Perhaps Lindemann pulled a few strings on that occasion too, despite the rather unpleasant encounter we had had a year before. He was after all a very decent man, and I am convinced that as his friend Winston's advisor in matters of physics he proved invaluable in the defence of Britain during the war.

The fifth period (1939-56) I shall call 'My Long Exile', but without the bitter associations of the word, as it was a wonderful time. I would never have got to know this remote and beautiful island otherwise. Nowhere else could we have lived through the Nazi war so untouched by problems that it is almost shameful. I can't imagine spending seventeen years in Graz 'treading water', with or without the Nazis, with or without the war. Sometimes we would quietly say amongst ourselves: 'Wir danken's unserem Fuhrer' ('We owe it to our Fuhrer').

The sixth period (1956-?) I shall call 'My Late Viennese Period'. As early as 1946 I had been offered an Austrian chair again. When I told de Valera about it he urgently advised me against it, pointing to the unsettled political situation in Central Europe. He was quite right in that respect. But while he was so kindly disposed towards me in many ways, he showed no concern for my wife's future should anything happen to me. All he could say was that he wasn't sure what would happen to his wife in such a situation either. So I told them in Vienna that I was keen on going back, but that I wanted to wait for matters to return to normal. I told them that because of the Nazis I had been forced to interrupt my work twice already and start all over again elsewhere; a third time would certainly put an end to it altogether.

Looking back, I can see that my decision was right. Poor Austria had been raped and was a sad place to live in those days. My petition addressed to the Austrian authorities for a pension for my wife as a kind of reparation was in vain in spite of the fact that they seemed keen to make amends. The poverty was too great then (and still is today in 1960, for that matter) to make allowances for certain individuals and deny them to almost all others. Thus I spent ten more years in Dublin, which turned out to be of great value to me. I wrote quite a number of short books in English (published by Cambridge University Press) and continued my studies on the 'asymmetric' general theory of gravitation, which appears to be disappointing. And last but not least there were the two successful operations in 1948 and 1949 by Mr Werner, who removed the cataracts from both my eyes. When the time had come, Austria very generously restored me to my former position. I also received a new appointment to Vienna University (extra status), although at my age I could only expect two and a half years in office. I owe all this mainly to my friend Hans Thirring, and to the Minister of Education, Dr Drimmel. At the same time my colleague Robracher successfully pushed the new law for the status of Professor Emeritus and thus also supported my cause.

This is where my chronological summary ends. I hope to add a few ideas or details here and there that are not too boring. I must refrain from drawing a complete picture of my life, as I am not good at telling stories; besides, I would have to leave out a very substantial part of this portrait, i.e. that dealing with my relationships with women. First of all it would no doubt kindle gossip, secondly it is hardly interesting enough for others, and last but not least I don't believe anyone can or may be truthful enough in those matters.

This summing-up was written early this year. It now gives me pleasure to read through it occasionally. But I have decided not to continue - there would be no point.